Lecture Notes in Computer Science 7100

Commenced Publication in 1973
Founding and Former Series Editors:
Gerhard Goos, Juris Hartmanis, and Jan van Leeuwen

W0192948

Abdelkader Hameurlain   Josef Küng
Roland Wagner (Eds.)

Transactions on Large-Scale Data- and Knowledge- Centered Systems V

 Springer

Editors-in-Chief

Abdelkader Hameurlain
Paul Sabatier University
Institut de Recherche en Informatique de Toulouse (IRIT)
118 route de Narbonne, 31062 Toulouse Cedex, France
E-mail: hameur@irit.fr

Josef Küng
Roland Wagner
University of Linz, FAW
Altenbergerstraße 69, 4040 Linz, Austria
E-mail: {jkueng,rrwagner}@faw.at

ISSN 0302-9743 (LNCS)   e-ISSN 1611-3349 (LNCS)
ISSN 1869-1994 (TLDKS)
ISBN 978-3-642-28147-1   ISBN 978-3-642-28148-8 (eBooK)
DOI 10.1007/978-3-642-28148-8
Springer Heidelberg Dordrecht London New York

Library of Congress Control Number: 2012930838

CR Subject Classification (1998): H.2, I.2.4, H.3, H.4, J.1, H.2.8

Typesetting: Camera-ready by author, data conversion by Scientific Publishing Services, Chennai, India

Printed on acid-free paper

Springer is part of Springer Science+Business Media (www.springer.com)

Preface

After the inaugural volume and three special-topic volumes in the areas of data warehouses and knowledge discovery, grid and P2P systems, and database systems in biomedical applications, respectively, this is the second regular volume of the Transactions on Large-Scale Data- and Knowledge-Centered Systems.

These contributions were submitted directly to the editors, who managed the traditional peer reviewing process. Recognized scientists checked the quality and gave helpful feedback to the authors – many thanks to them from the editorial board. As a result we find four general topics within this volume: query processing, information extraction, management of dataspaces and content, and mobile applications.

The first part is the largest one. It consists of four contributions that deal with several aspects of query processing. In *"Approximate Query Processing for Database Flexible Querying with Aggregates"* the authors describe the use of formal concept analysis together with database techniques, enabling aggregate queries to be executed efficiently but at the cost of some imprecision in the results. The second paper *"Metric-Based Similarity Search in Unstructured Peer-to-Peer Systems"* introduces a P2P system for similarity search in metric spaces by applying an interconnected M-tree structure. *"Adaptive Parallelization of Queries to Data Providing Web Service Operations"* extends a given web service mediator system with a particular operator that supports a substantial performance improvement. And finally, *"A Pattern-Based Approach for Efficient Query Processing over RDF Data"* presents a solution that uses particular indices of structural patterns to speed up such queries on data sets published in RDF format.

Information extraction is the focus of the second part. *"The HiLeX System for Semantic Information Extraction"* describes a new ontology-driven approach that aims to increase the recall rate, and which has been employed in real-world applications with positive feedback.

The third cluster consists of two articles. *"DSToolkit: An Architecture for Flexible Dataspace Management"* presents a toolkit for managing the entire life cycle of a dataspace. It supports the integration of large numbers of evolving data sources and on-the-fly data integration. The second paper on this topic, *"Temporal Content Management and Website Modeling: Putting Them Together"*, presents a methodology for the design and development of temporal and data-intensive websites by using three particular models: a conceptual model, a navigation scheme, and a logical schema.

'Mobile aspects' is the common denominator for the final part. The authors of *"Homogeneous and Heterogeneous Distributed Classification for Pocket Data Mining"* present an adaption of data stream classification techniques for mobile and ad hoc distributed environments, and finally *"Integrated Distributed/Mobile Logistics Management"* describes an architecture within which it is possible to integrate logistics management systems of different cooperating transport companies or mobile users.

Last but not least we would like to thank Gabriela Wagner for supporting us with the organization and we hope that you enjoy this TLDKS volume.

November 2011

Abdelkader Hameurlein
Josef Küng
Roland Wagner

Editorial Board

Table of Contents

Approximate Query Processing
for Database Flexible Querying with Aggregates

Minyar Sassi[1], Oussama Tlili[2], and Habib Ounelli[2]

[1] National Engineering School of Tunis,
BP. 37, le Belvédère 1002 Tunis, Tunisia
minyar.sassi@enit.rnu.tn
[2] Faculty of Sciences of Tunis,
Campus Universitaire, 1060 Tunis, Tunisia
{oussama.tlili,habib.ounelli}@fst.rnu.tn

Abstract. Database flexible querying is an alternative to the classic one for users. The use of Formal Concepts Analysis (FCA) makes it possible to turn approximate answers that those turned over by a classic DataBase Management System (DBMS). Some applications do not need exact answers. However, database flexible querying can be expensive in response time. This time is more significant when the flexible querying require the calculation of aggregate functions ("Sum", "Avg", "Count", "Var" etc.). Online aggregation enables a user to issue an SQL aggregate query, see results immediately, and adjust the processing as the query runs. In this case, the user sees refining estimates of the final aggregation results. In spite of the success which known this method, until now, it hasn't been integrated in flexible querying systems. In this article, we propose an approach which tries to solve this problem by using approximate query processing (AQP). This approach allow user to i) wrote flexible query contains linguistics terms, ii) observe the progress of their aggregation queries and iii) control execution on the fly. We report on an initial implementation of online aggregation in a flexible querying system.

Keywords: Flexible Querying, Aggregate Functions, Fuzzy Logic, Approximate Query Processing, Formal Concept Analysis.

1 Introduction

Classic querying systems make it possible to represent and store data in a precise way in order to find them by using strict Boolean conditions.

In this case, the user must know the exact data organization to express his query. It is one of the limits of the classic querying.

Database (DB) flexible querying aims at improving the capacity of expression of the query languages in order to take into account preferences expressed by users and to facilitate the access to the relevant information not represented explicitly in the database [1,2].

Thus, a result answer will be more or less relevant according to the user preferences.

A. Hameurlain et al. (Eds.): TLDKS V, LNCS 7100, pp. 1–27, 2012.

Several works was proposed in the literature to introduce the flexibility into the DB querying [2,3,4,5,6,7,8,9,10,11,12]. The majority of these used the fuzzy set formalism to model linguistic terms such as "small" or "young".

The use of the concepts lattice, core of Formal Concepts Analysis (FCA) [13], in querying gives hierarchical answers in several levels.

In addition, the decision-making and Data Mining [14] applications have often recourse to the aggregate functions like "Sum", "Avg", "Count", "Var" etc. in the query formulation. These functions are carried out on large DB what spends much time. For these type of query, the precision is not required, thus a user wishes to know as soon as possible an approximate response instead of waiting several minutes to have the exact answer.

Approximate Queries (AQ) [15,16,17] use techniques which support the calculating time compared to the exactitude of the answer.

With AQ, we can gradually have answers approached until leading to the exact response at the end of the calculation.

The goal of this work is to propose an intelligent DB flexible querying approach while combining FCA and AQ making it possible to provide gradually, of the approximate answers with the maximum of confidence rate in an acceptable time.

Our approach tries to minimize the query response time while maximizing the confidence rate of obtained answers.

The rest of this article is organized as follows: in section 2, we present a state of the art of the recently database flexible querying systems as well as AQ techniques. Section 3 presents the proposed DB flexible querying with aggregates. In section 4, we carry out experiments in order to evaluate the proposed approach. Finally, section 5 concludes the work and gives some future works.

2   State of the Art

We present in this section, a state of the art the recent works on database flexible querying and AQ.

In these works, flexibility was interpreted in a different way. On the one hand, to facilitate as much as possible the query expression in terms of precision and, on the other hand to make the query expression closer to the natural language. The second interpretation relates to the query evaluation process.

The classic querying systems distinguish two categories of data: those which satisfy the search keys and those which do not satisfy them. The principle of the flexible querying flexible aims at extending this binary behavior by introducing the graduality concept.

The work undertaken for AQ concentrated on the sampling techniques [17], Wavelets [15] or Histograms [16]. We limit ourselves to those based on sampling which are closer to our work.

2.1   Flexible Querying

The majority of the works undertaken in the field of the DB flexible querying make the assumption that a query has an aim the search of the data satisfying a Boolean condition.

Indeed problems appear when the user don't express himself directly like a Boolean condition (in all or nothing), but is on the contrary gradual, because of the existence of preferences on the level of these search keys [2]. These problems can be solved by a flexible querying.

Flexible Query. A flexible query is a one which comprises vague descriptions and/or vague terms which can comprise preferences [2].

Let us consider a user who consults a DB of car's hiring offers. The user wishes to find a car preferably with *Red Colour,* an "accessible" Price and a *"small"* Seniority. The user wishes write his query without taking as follows:

"Select * From Cars Where Colour red And Price accessible and Seniority small"

Several works were proposed in the literature to introduce the flexibility into the DB querying such as the use of the complementary criteria, the preferences, the distance and the similarity, models based on the fuzzy set theory and recently approaches based on the Type Abstraction Hierarchies (TAH) and Multi-Attributes Type Abstraction Hierarchies (MTAH) [9,10], approaches based on FCA [11]and those on a FCA fuzzification[12].

The TAH, firstly proposed in [9] makes it possible to represent the DB fields on several abstraction ones.

The higher levels of the hierarchy provide a more abstract data representation than the lower levels. Generalization (ascending the hierarchy), specialization (descending the hierarchy) and association (moving between the hierarchies) are the three key operations in the fact of drawing from the cooperative answers for the user.

The contributions of this approach are:

– The use of TAH and MTAH structures to model generalization and specialization by concepts hierarchies.
– The taking into account of the semantic dependences between the query search keys to determine its realisability or not. This constitutes a form of optimization of the flexible and Boolean queries.
– Cooperation with the user while returning answer closer to the initial query by a generalization process in the case of non contradictory query but having an empty answer.

However, this approach presents the following limits:
– The non detection of the query realisability.
– The generation of the false answers.
– Users must also know the DB organization since they must specify the attributes which they must release or not as well as the level of relieving of each attribute.

In [10], an approach of relieving of a classic query which uses predicates with relieving attributes, i.e. attributes which the users can use in a comparison predicate with a linguistic term like *"average"* Price instead of *"Price between 100 and 300"*.

For example, the attribute "Price" is relievable and the users can refer it by one of the linguistic terms *"weak"*, *"average"*, and *"enormous"*. The clusters, generated for each relievable attribute, are stored in the catalogue of the DBMS, in the form of particular tables called BCAR. This partitioning is carried out only once for each relievable attribute.

Fig. 1 shows the general principle of this approach [10]:

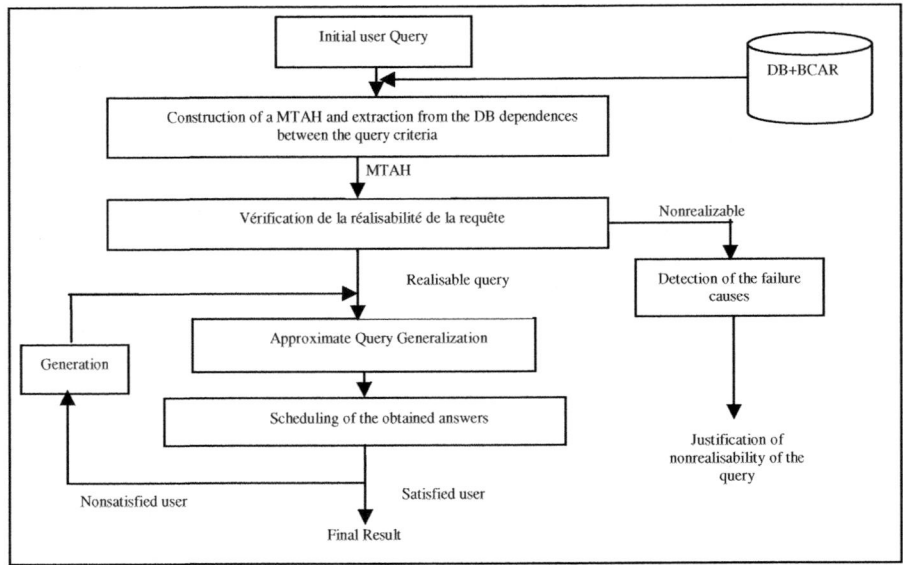

Fig. 1. Flexible and cooperative querying by hierarchical conceptual abstraction

The contributions of this approach are:

- No modification of SQL, the user is not obliged to make random and repetitive choices, as it is the case in several systems such as FLEX [5], VAGUENESS [7] and CoBase [9].
- The relievable attributes are fixed by the DB Administrator (DBA). This is all the more significant since it is addressed to end-users not having the precise knowledge on the data organization. It is easier to an expert to specify which attribute is relievable and than it can be used with linguistic terms. This facility is more interesting than to use *Within (100,120,150,300)* operator of CoBase.

However, this approach present some limits at the level of the structures which it uses. We can quote:

- The incremental maintenance of Knowledge Base (KB) of the relievable attributes (BCAR),
- The clustering of the relievable attributes without fixing, a priori, the number of clusters to be generated,
- The problem of storage of the clusters and indexing of the MTAH.

In [11] a flexible querying approach based on FCA [13] was proposed. It makes it possible to integrate the concepts lattice, core of FCA, in the flexible DB querying. It aims to give approximate answers in the event of empty answer and to evaluate a flexible query by determining the failure causes. It gives others realizable queries with approximate answers while being based on the concepts lattices.

The steps of this approach are described in Fig. 2 [11]:

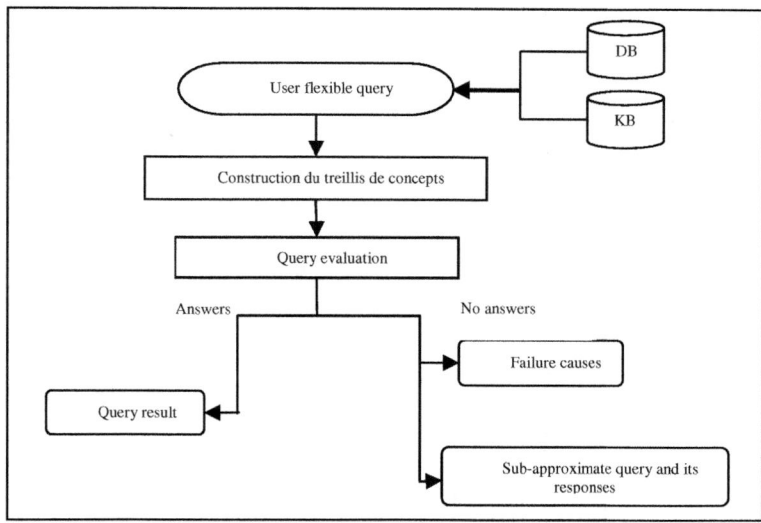

Fig. 2. Flexible querying based on FCA.

The contributions of this approach are:

− The processing of the case of empty answers for a flexible query
− The detection of the failure causes of a flexible query
− The generation of the approximate sub-queries likes their answers.
− The user will have an indication on the incompatible criteria allowing him to reformulate other queries having answers. This minimizes the formulation of queries generating an empty answer.
− The no modification of the structure of SQL language and thus to profit from the functionalities from the DBMS.

However, this approach presents limits such as:

− The use of the linguistic terms such as *"very"*, *"enough"*... in the query. This criterion is interesting since the users use these linguistic terms in general.
− This system present some limits when the DB is large. This is due to the cost of the construction of concepts lattice associated with the user query.

In [12] a DB flexible querying approach based on a fuzzification of concepts lattices was proposed. The fuzzification is the process of conversion of a precise data into fuzzy data which will be represented by one or more fuzzy sets described by membership functions.

The formal context instead of containing data binary (0 or 1), it contains the membership degrees (in the interval [0,1]).

The general principle of this approach [12] consists in organizing data in order to optimize querying. The concept of query was introduced to check the query satisfaisability. The turned over answers were classified according to a satisfaction degree calculated compared to the requirements specified in the initial query.

It does not require modifying SQL to querying data. Searches keys in a query, flexible or not, are not necessarily independent.

This approach is based on an extension of the MTAH structure. The querying process proceeds in two dependent steps. Fig. 3 represents the principle of this approach [12].

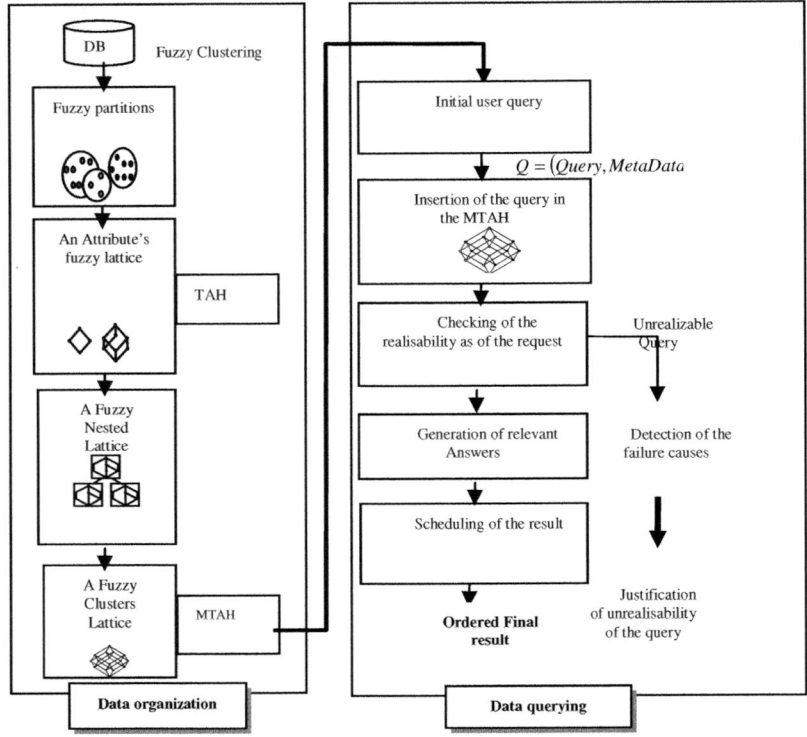

Fig. 3. Flexible querying based on Fuzzification of FCA [12]

The contributions of this approach are:
- The generated clusters for each relievable attribute are not stored any more in the catalogue of the DBMS.
- The number of cluster is automatically determined.

However, this approach presents limits such as:
- The sensitivity of the quality measures, used in the determination of the optimal number of clusters, to the phenomenon of cluster's overlapping.
- The clustering is a central task in the organization step. This problem becomes complex when it is about large DB.

2.2   Approximate Query

During last years, progress in the data-gathering led to a proliferation of very large DB such as DataWarehouse. In practice, however, whereas the collection and the massive data storage became relatively simple, querying of such volume is more difficult. For large DB, that results in an access to enormous quantities of data, which generates an enormous cost in execution time.

This influences considerably the feasibility of many types of analysis applications, in particular those which depend on the topicality or the interactivity.

The query response time is very significant in many decision-making applications. However exactitude in the query results is less significant. For example, to know the marginal data distribution for each attribute up to 10% of error can largely be enough to identify the sale of products in sale's DB.

Of this effect, AQ contribute for an effective solution which sacrifice exactitude to improve the response time.

It is used in the aggregate query (comprising "Sum", "Avg", "Count", "Var" etc.) in which the precision with the "last decimal" is not very asked.

AQ uses several techniques with knowing:

- Histograms [16]
- Wavelets [15]
- Sampling [17]

We use the sampling which seems to us most suitable for the flexible querying which interests us in this work.

It consists in building tables (or views) by selecting some tuples starting from the initial table to build a sample. This one has a less significant storage size than the initial table.

Instead of querying all the DB, we querying a sample which represents the DB and we obtain then approximate answers of a query as shown in Fig. 4.

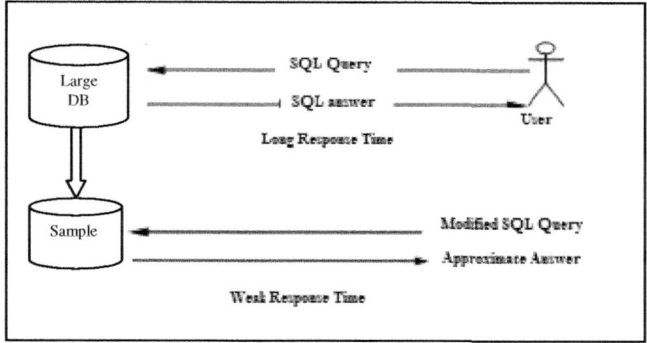

Fig. 4. Approximate Query Execution

In spite of their efficiency in the generation of the approximate answers and the aggregate queries processing, flexible querying approaches as well as the AQ presented previously present the following limits:

- No taking into account of the aggregate queries: *"Average Salary of the Young employees"*,
- The exactitude of the answer and,
- The time of the answer put for the construction of the final answers.

For aggregate queries, we propose in this article a data course path while using FCA in order to generate a hierarchy allowing the user to personalize these answers in several abstraction levels.

For the exactitude of the answer, we propose to use AQP (Approximate Query Processing) which consists of techniques which sacrifice the exactitude in order to improve the response time.

In order to improve the response time, we propose to adapt aggregation on line technique proposed in [21] whose objective is to gradually give approximate answers at the same time of query execution.

It consists to apply a sampling to the initial data of the DB in order to minimize the disc access to the disc and consequently to improve the response time.

In section 4, we present a new DB flexible querying approach combining AQ and FCA while taking into account aggregate queries.

3   Flexible Querying with Aggregate

In this section, we present a new DB flexible querying approach while combining FCA and AQP while taking into account aggregate queries.

3.1   System Architecture

In this sub-section, we present the querying system architecture. It is presents in Fig. 5. It comprises the following components:

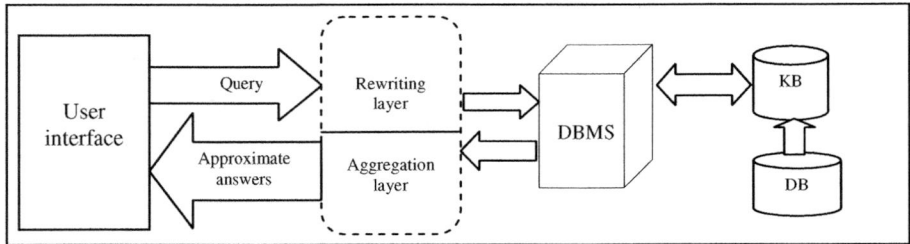

Fig. 5. Architecture of proposed querying system

- Rewritable layer: she takes care of rewriting the aggregate query in its final form by adding aggregate functions and calculating the Error Rate depending on the confidence degree defined by user. The query becomes an AQ.
- Aggregation layer: it is responsible for transferring the user with different answers gradually during the query execution. It gives the Error Rate.
- DB: it is a relational DB where we store all permanent information in a relational model.
- KB: it is a Knowledge Base that is generated from the DB and before the query execution. It contains information on the relievable attributes (an attribute that describes a linguistic term). It's schema is described in Table 1.

Table 1. KB Schema

ID row	Relievable-Attribute1	Relievable-Attribute 2	···	Relievable-Attribute n
...	....	....	....	......

3.2   Detailed Description of Approach Steps

Our approach is described in Fig. 6. It is divided into two major steps:

- Pre processing step: in this step, we will generate the KB from the DB which contains membership degrees of each tuple to each relievable attributes.
- Post processing step: when the user launches the application, the system searches for approximate answers, and then calculates the aggregation and gradually sends to the user.

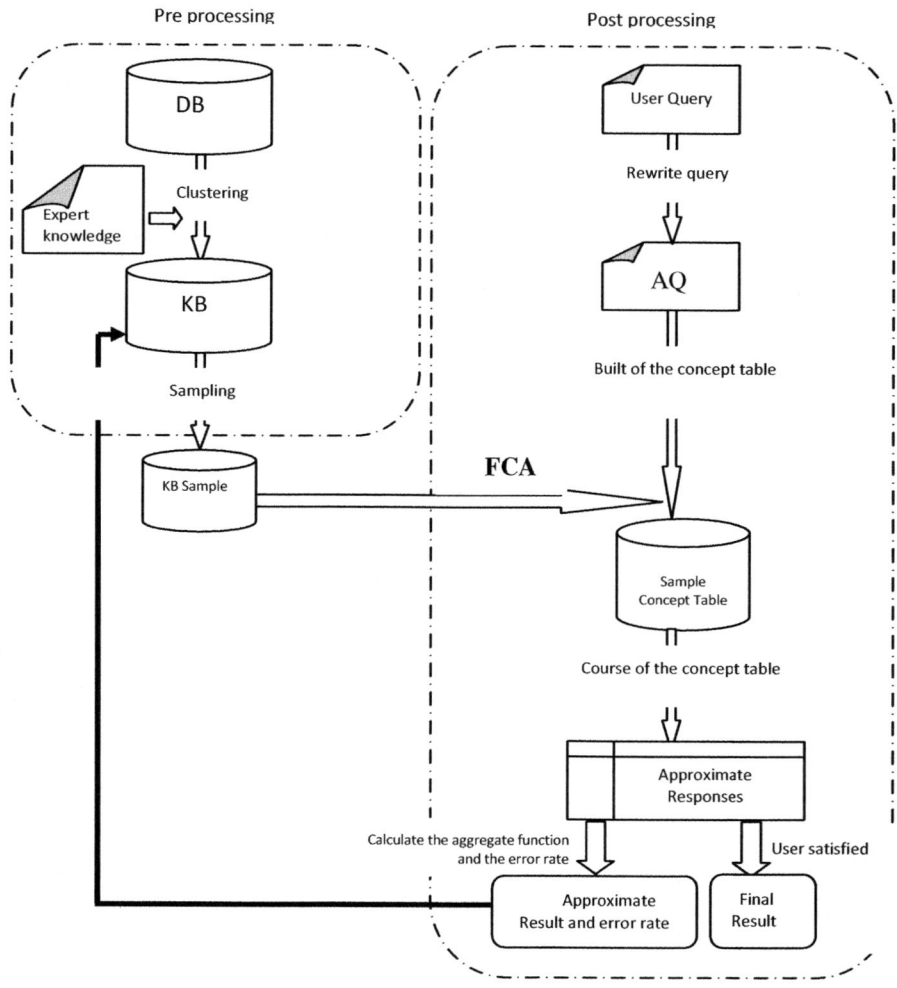

Fig. 6. The Approach Steps

3.2.1 KB Construction

Before query execution, we generate a KB which contains membership degrees of each tuple for each relievable attribute by using membership function. Zadeh proposes a series of membership functions [19]. We quote essentially Triangular function, Singleton function, LFonction function, Gamma function, Trapezoidal function….

We will use the Trapezoidal function; it is given by the equation (1).

$$\mu_E(x) = \begin{cases} 1 & if\ C \leq X \leq B \\ \dfrac{X-A}{B-A} & if\ A \leq X \leq B \\ \dfrac{D-X}{D-C} & if\ C \leq X \leq D \\ 0 & if\ X \prec A\ or\ X \succ D \end{cases} \tag{1}$$

Example. We present in Table 2 data membership degrees for the relievable attribute *"Age"* generated by the Trapezoidal function.

Table 2. Membership degrees of "Age" attribute

Id_tuple	Age_Young	Age_Adult
25	0.7	0.3
30	0.2	0.8
20	1	0

If *Age_Young=0.7* for the first tuple, then this last has a membership degree equal to 0.7 for the group *Age_Young*.

3.2.2 Flexible Query Rewriting

The first step of query execution consists of construct the AQ through an interface in which it specifies the confidence degree, the concept table, the aggregate function ("Sum", "Avg", "Count", "Var" etc.), all attributes of the Select and Where clauses.

In this article, we consider a single table without using Group By clause that it contains thousands of records. The AQ is as follows:

Select function(attribute), confiance_degree as confidenc ,

functionInterval(confiance_degree) from table   where attribuet1   Is flexible_ condition1 [and … attribute2 Is flexible_ condition2]

where function( ) and functionInterval( ) are user predefined functions and which can give online approximate answers depending on the confidence degree for aggregate "Sum", "Avg", "Count", "Var" etc.

3.2.3 Data Sampling

Instead of querying the entire KB, we interview a sample of KB that is made up of hundreds of records which improves response time.

DB sampling algorithms were proposed in the literature, such as Olken algorithm [17]. The basic idea of this algorithm is to carry out the sampling of a table without going to physically read all the lines of the latter. Sampling is done using Oracle function called "Sample" [18]. It allows a simple random sampling without handing-over by selecting randomly recordings of a manner to choose recording only once. This function has like parameter p the percentage of sampling table.

We repeat the construction of the sample at the same execution query time by taking account of the following conditions:

- The percentage of sampling is fixed by the DBA.
- The recordings chosen in the preceding samples are not selected any more.

We finish the construction of the samples until all the recordings were chosen or the user finished the program execution.

The sampling procedure is described as follows:

```
Procedure Sampling
Inputs :     Q : Query
             KB: Knowledge Base
             p : Sampling percentage
                        n : KB size
Output : E: Sample
Begin
  Step 1 : KB1 := All tuples starting from KB which contain only the
columns which belongs to Q.
  Step 2 : E contains  n*p/100  tuples released randomly from KB1 which do
not belong to E.
  Step 3 : Repeat steps  1 and 2 Until all treated tuples.
End.
```

3.2.4 Building Sample Concepts Table

The concepts table [20] is a tabular representation of a concept lattice, core of FCA, and its construction is easier than the lattice.

It is described in Table 3 where the columns have the following meanings:

- *C#* (context#): the name of the source context.
- *Niv#, N#* (Level#, Node#) :these two columns store the identifier of the concept. The first is the level of the concept in the lattice while the second represents the sequence number of the concept at this level.
- *Int#, Ext#* (Intention, Extension): these columns store intensions and extensions of each concept.
- *L_s#,L_p#(*Successors List, Predecessors List): these two columns store the identifiers of successors (predecessors respectively) of the concept.
- *T_i,T_e* (Size_Intension, Size_Extension): these two columns store the cardinality of a concept (respectively the number of attributes and the number of tuples).

Table 3. Sample concepts table

C#	Niv#	N#	Int#	Ext#	L_s#	L_p#	T_i	T_e
.....	....	...	...	...	....	....	....	....

3.2.5 Coursing the Sample Concepts Table and Calculating Aggregation

In this step, we course the sample concepts table to extract approximate answers and to calculate the final result of AQ.

We use the proposed algorithm in [20] to build a sample concepts table on the AQ and then return the approximate answers. In order to improve the response time, we build the concepts table using only the query conditions. This reduces the table size and minimizes the complexity of the construction of the sample concepts table.

In the case of empty answers, we traverse the concepts table to release from other approximate answers by calculating the values of the concepts of the concerned higher level.

The calculation of each aggregate function is started with the following descriptions:

– *Value(t)* : represents the aggregate value of the tuple *t*.
– *Degree(t)* : represents the membership degree of *t*.

Calculating_Sum, Calculating_Count, Calculating_Count and *Calculating_Avg* algorithms make it possible to calculate *Sum, Count* and *Avg* functions of all attributes while taking into account of membership degrees. These algorithms take in input a concepts table *TCX*, the maximum *Max* and minimum *Min* values of attributes to be incorporated and finally the sample size *Size*. These values are used for calculation of the *Error Rate*.

```
Algorithm   Calculating_Sum
Inputs:     Concepts  Table: TCX
                      Maximum value of attribute: Max
                      Minimum value of   attribute: Min
                      Sample Size :n
Outputs :   Result : Res
                      Error Rate : Rate
Begin
            D=1,
            Sum=0
            For  Each  Element E of the concepts  table TCX

                If concept extension ≠∅      then
                             For each tuple t of the concept extension
                                  Sum=Sum+Value(t)
                                      If Degree(t)< D Then  D=Degree(t)
                             End For
                End If

            End For
```
$$Res = Sum * D$$
$$Rate = \frac{1.22 * (Max - Min)}{\sqrt{n}}$$
```
End.
```

```
Algorithm Calculating_Count
```

Inputs:	Concepts Table: *TCX*
	Maximum value of attribute: *Max*
	Minimum value of attribute: *Min*
	Sample Size :*n*
Outputs :	Result : *Res*
	Error Rate : *Rate*

Begin

      D=1

      Card=0

    For Each Element *E* of the concepts   table *TCX*

        If concept extension $\neq \emptyset$ then

            For each tuple *t* of the concept extension

               card=card+1

          If *Degree(t)< D* **Then** *D=Degree(t)*

            End For

        End If

  End For

  *Res= Card * D*

$$Rate = \frac{1.22*(Max-Min)}{\sqrt{n}}$$

End.

```
Algorithm Calculating_Avg
```

Inputs:	Concepts Table: *TCX*
	Maximum value of attribute: *Max*
	Minimum value of attribute: *Min*
	Sample Size :*n*
Outputs :	Result : *Res*
	Error Rate : *Rate*

Begin

      D=1

     Card=0

    Sum=0

    For Each Element *E* of the concepts   table *TCX*

        If concept extension $\neq \emptyset$ then

            For each tuple *t* of the concept extension

              Sum=Sum+Value(t)

              Card=Card+1

          If *Degree(t)< D* **Then** *D=Degree(t)*

            End For

        End If

        End For

$$Res = \left(\frac{Sum}{Card}\right)*D$$

$$Rate = \frac{1.22*(Max-Min)}{\sqrt{n}}$$

End.

To calculate the aggregation we perform aggregate functions predefined by the DBA [21]:

− For Avg( ) function :

$$Avg = \left(\frac{1}{n}\right)\sum_{i=1}^{n} v(L_i) * \deg ree \qquad (2)$$

− For Sum( ) function:

$$Sum = \sum_{i=1}^{n} v(L_i) * \deg ree \qquad (3)$$

− For *Count( ) function:*

$$Count = \sum_{i=1}^{n} 1 * \deg ree \qquad (4)$$

where $\deg ree = Min(U_{i1} \wedge^\wedge V_{i1} \wedge ... \wedge Z_{i1})$ and U, V, Z are the membership degrees on the query Q and n is the sample size, $v(L_i)$ is the value of the tuple index i (L_i is a random index).

We calculate the Error Rate (Interval) associated with the aggregate function. We use the method of conservative confidence intervals [11]:

$$(b-a)\left(\frac{1}{2n}\ln\left(\frac{2}{1-p}\right)\right)^{\frac{1}{2}} \qquad (5)$$

where [a, b] is a predetermined interval, such that $a \le v(i) \le b$ for all $1 \le i \le m$, n is the sample size, m is the size of KB, p is the confidence rate (example $p = 0.95$)

3.3   Illustrative Example

Let a simple relational DB table *employee (Id, Name, Age, Salary)* which contains tuples given in table 4.

Table 4. Example of a relational DB table "employees"

ID_Tuple	Name	Age	Salary
1	Mohamed	23	400
2	Ali	30	550
3	Walid	45	700
....	....	....	.....
10000	Wajdi	40	800

The relievable attributes are *Age-Young, Age-Adult, Salary-Low, Salary-Middle, Salary-High*. KB is given in table 5.

Table 5. KB

ID_Tuple	Age_Young	Age_Adult	Salary_Low	Salary_Middle	Salary_High
1	0.7	0.3	0.6	0.4	0
2	0.5	0.5	0	1	0
3	0	1	0.1	0.6	0.3
....	....	...	...	...	...
10000	0.1	0.9	0	0.3	0.7

Then, we eliminate data with low membership degrees by setting a user defined threshold. KB becomes as shown in Table 6:

Table 6. KB with a threshold

ID_Tuple	Age_Young	Age_Adult	Salary_Low	Salary_Middle	Salary_High
1	0.7	-	0.6	0.4	-
2	0.5	0.5	-	1	-
3	-	1	-	0.6	-
....	....	...	...	...	...
10000	-	-	0.9	-	0.7

Consider the following query *"Average (Avg) Salary for Young employees having Low Salary with a confidence rate = 95%"*.

Select Avg(Salary) from emploee where age Is Young and Salary Is Low

The AQ becomes:

Select AVG(Salary), 0.95 as confidence , ConsAvgInterval(0.95) from employee where age Is Young and Salary Is Low

We construct the sample (Table7) according to the KB at the same time of query execution.

Table 7. Sample of data

ID_Tuple	Age-Young	Salary-Low	Salary
1	-	1	400
20	0,8	-	900
520	-	0.9	430
32	-	0.8	460
10	0.6	-	780
...	....	....	....
130	-	0.5	550

Then we generate a concepts table associated with the query as shown in Table 8:

Table 8. Sample concepts table

C#	Niv #	N#	Int#	Ext#	L_#	L_p#
1	1	1	Young_A,	ϕ	(1,2,1)	0
			Low_Salary		(1,2,2)	
1	2	1	Low_Salary	1(1 ;400), 32(0.8 ;460),	(1,3,1)	(1,1,1)
				520(0.9;430),		
				130(0.5;550)		
1	2	2	Young_Age	10(0.6 ;780), 20(0.8;900),	(1,3,1)	(1,1,1)
1	3	1	ϕ	1(1 ;400), 32(0.8;460),	0	(1,2,1)
				520(0.9;430),		(1,2,2)
				,130(0.5;550),		
				10(0.6 ;780), 20(0.8;900)		

Each extension contains two parameters: the degree and the aggregate value.

Example. 20 (1, 380) the tuple number 20, a degree is 1and its aggregate value is 380.

We note that we do not have any response for *"Young_Age"* and *"Low_Salary"*. The system turns over the result of the higher levels and does not turn over 0 as shown in Table 9.

We repeat these steps until all the KB is treated either we get an Error Rate is very low to say the exact result is very close to either the user is satisfied with the outcome and conclusion the query execution.

In Table 9, we present an example of turned results after the calculation the average Avg and the *Error Rate* functions.

Table 9. Approximate Answers

Avg	Attribute	Confiance	Error Rate
400	Low_Salary	95 %	0.06503
402	Young_Age	95%	0.06500
405	Low_Salary	95%	0.06470
....		...	......
410	Young_Age	95%	0.0090

In this section, we presented a DB flexible querying approach allowing integrating AQ in a system using FCA in order to raise the limits found in the preceding approaches when we use aggregate queries.

In the following section, we will present experimentation and evaluation of the proposed approach.

4   Experimentation and Evaluation

After presented the theoretical framework of our work, we present in this section experimentation and evaluation of the proposed approach. We start by presenting the implementation of the different steps, then, we carry out to evaluate it. We approach the experiments to evaluate the performances of this approach while basing ourselves over the response time and the exactitude of turned approximate answers.

4.1   Implementation of the Approach Steps

We implemented our approach with the Java language while using Oracle 10g DBMS.

4.1.1 Pre Processing Step

This step consists in building the KB starting from the DB. A domain expert knows the various attributes of DB and classifies the data in various groups or clusters.

The domain expert stores in a table called Expert Table , as indicated in Table 10, the parameters A, B, C, D of the Trapezoidal function of each relievable attribute to be able to determine the membership degrees.

Consider the example defined in the preceding section. The table built by the expert is as follows:

Table 10. Expert Table

Relievable_Attribute	A	B	C	D
Low_Salary	0	0	400	500
Middle_Salary	400	500	700	800
High_Salary	700	800	1200	1400

The KB generation procedure is described below. It takes as inputs, the Expert Table EXT which contains the values of the Trapezoidal function and the DB table DBT.

We traverse EXT tuple by tuple, and for each tuple we calculate the membership degree (degree of each tuple T of DBT according to the concerned attribute value and (A,B,C,D) values. Finally we add the membership degree to the KB.

```
Procedure   KB Generation
Inputs :    Database table: DBT
            Expert Table: EXT
Output :    Knowledge Base: KB

Begin
  For Each element E of EXT Do
        ATT = Value(Relievable_Attribute)
        Add a new Attribute ATT to KB
        For each tuple T of DBT Do
            degree= function_trapesoidal(T, ATT)
              Add val to KB
        End For
  End For
End.
```

4.1.2 Post Processing Step

In this step, the user provides the following parameters:

- The confidence degree which is useful for calculation of the Error Rate or the interval.
- The threshold: the minimum value which is used in order to eliminate the answers which have a small membership degree.
- The aggregation function: "Sum", "Avg", "Count", "Var" etc.
- The attribute to be incorporated,
- The source DB table (DBT) ;
- Flexible query conditions.

Answers are progressively displayed in the data grid in the same time of the query execution. In this case, the user is not obliged to await the end of the execution to have the final result. Indeed, after a certain time, the first answer and its Error Rate are displayed, until the user stops calculation or which the KB was entirely treated.

The answer is composed of two parts:

- Approximate answers.
- Error Rates (the interval in which is the exact answer).
    Example. Answer 1= 585.88 Dt   + /- 16,682   with 95 % Confidence
               Answer 2 = 613.11 Dt +/- 11,477   with 95 % Confidence

4.2   Evaluation

In this sub-section, we approach the experiments carried out to evaluate the performances of our DB flexible querying approach while basing ourselves over the response time and the exactitude of the result answers.

4.2.1 About Response Time

We tested the performances of our system using several DB

- Test with « employees » DB table

We started with the DB table *employees (Id, Name, Salary, Age)* and we increased the number of tuple from 789 to 9498. We calculate for each case the response time given by our system and compare it with the case of a classic querying (without AQ) as shown in Fig. 7.

This table contains two relievable attributes relaxables *"Age"* and *"Salary"*.

The query is as follows *"The average of the weak Salary of the young employees"*

The query is as follows: *"Select Avg(Salary) From employee Where Salary Is weak And Age Is young"*

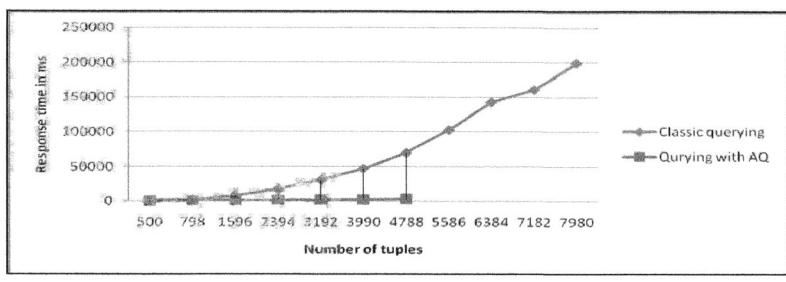

Fig. 7. Comparison between Classic Querying and querying with AQP via "Employees" DB table

According to Fig. 7, we note that the response time is weaker by using AQ than that without AQ.

For the classic querying (without AQ), the curve is exponential, whereas for the querying with AQ, the curve is ~ linear.

If the size of the DB exceeds 7000 tuples, the response time, without AQ, is about 2 minutes, whereas it is, with AQ, about 5 seconds.

–   Test with *"Adult"* benchmark DB

The experiment was undertaken by using a benchmark DB called *"Adult"*. It is extracted starting from the UCI Machine Learning Repository [22].

The KB contains three relievable attributes: *"Age"*, *"Job"* and *"Richness"*.

The query is as follows: *"Select Avg(Age) From Adult Where Job Is hard And Age Is Young"*

Table 11. Comparison between Classic Querying and querying with AQP via "Adult" benchmark DB

Approach	Traditionnel Querying	Quarying with AQ
Number of tuple	77947	77947
Response time	529 seconds ≈ 8 minutes	3 seconds

As shown in Table 11, the response time emits with AQ is weaker than that without AQ.

4.2.2 About Response Exactitude

–   Test with « employees » DB table

To test the exactitude of the answer, we carry out the application in the classic querying case (without the use of AQ) and we await the final answer.

For the DB table "employees", the exact answer to have when we carry out the following query *"Select Avg(Salary) From employees Where Salary Is weak And Age Is young"* is 595.494 Dt.

Fig. 8 presents the evolution of the approximate answers until leading to the exact answer:

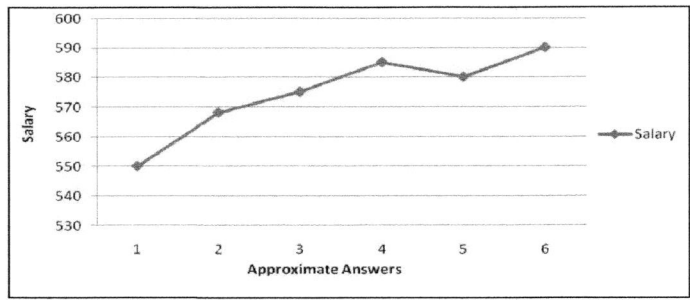

Fig. 8. Evolution of the approximate answers via "Employees" DB table

We note that the approximate answers improve more and more. The fourth answer (585 Dt) is very closer to the exact answer (595.494 Dt).

−   Test with "Adult" benchmark DB

We will carry out the application on the *"Adult"* benchmark DB with AQ and without AQ, then to check the quality of the result answers with AQ.

The exact answer to have when we carry out the following query *"Select Avg(age) From Adult Where Job Is middle And Age Is adult"* is 34.9 year.

Fig. 9 presents the evolution of the approximate answers until leading to the exact one:

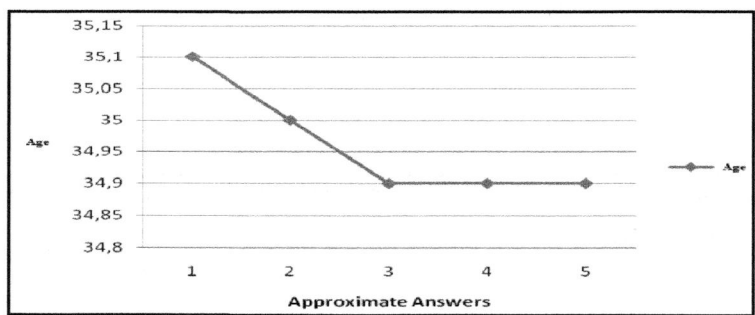

Fig. 9. Evolution of the approximate answers via "Adult" benchmark DB

We note that the approximate answers are very close to the exact one.

The 1st approximate answer is about 35.1 year + / - 1.2   with 95 % confidence degree whereas the exact answer (without AQ) is 34.9 year.

4.3   Complexity Study

4.3.1 Temporal Complexity

Let p the considered number of tuples in calculation compared to the total number of tuples N of *KB*, k is the number of relievable attributes. For each of two steps, temporal complexity is as follows:

- Pre processing step: the complexity is, in the worst case, is that of the *KB* construction procedure : $o(k \times N)$
- Post processing step:

For sampling procedure, the complexity *CT1* is about:

$$CT1 = O\left(\frac{1}{\frac{p}{100}}\right) = O\left(\frac{100}{p}\right) \tag{6}$$

For sample concepts table construction algorithm, the temporal complexity *CT2* is about:

$$CT2 = O\left(|G|^2\right) \tag{7}$$

Where $|G|$ is the row's number in the concepts table.

For coursing the sample concepts table and calculating aggregation, the temporal complexity *CT3* is about:

$$CT3 = O\left(|G| \times t\right) \tag{8}$$

where t is the size of the concept extension.

Then temporal complexity *CT* for the Post processing step is about:

$$CT = CT1 \times (CT2 + CT3) = \frac{100}{p} \times \left(|G|^2 + \left(|G| \times t\right)\right) \tag{9}$$

$$CT = O\left(\frac{100}{p}|G|^2\right) \tag{10}$$

4.3.2 Spatial Complexity

The proposed querying system use a sample concepts table in the shape of a matrix of m lines and n columns, it uses also a matrix for the sample of k lines and p columns.

The system thus holds $(n \times m) + (k \times p)$ box memory.

4.4   Comparative Study

In this section, we present the essential idea of the main DB flexible querying approaches closest to our. They differ mainly by the way used to find the values closest to those queried by the user and the used formalism to model uncertainty and imperfection of the real world.

The contributions of approaches proposed in [9,10] are important, including the concepts of TAH and MTAH for modeling generalization and specialization hierarchies of concepts. In this approach, no modification of SQL is required, what constitutes an asset for the implementation of this approach.

In this approach, the relievable attributes are set by the DBA. This is especially important that the proposed approach is aimed at end users with no specific and detailed knowledge on the data organization. It is easier for an expert to specify if an attribute is relievable or not and which linguistic terms we can use.

However, this approach has limitations in the structures it uses. We mainly include:

- The incremental maintenance of the KB relievable attributes,
- The clustering of relievable attributes without fixing a priori the number of clusters,
- The problem of storage ,clustering and indexing MTAH, and
- The not taking into account aggregate queries.

The approach proposed in [11] allows the treatment of empty response to a flexible query. Thus, it detects the failure causes and allows the generation of sub-queries and approximate answers.

Another advantage of this approach is that not changing the structure of SQL and thus benefit from the features of the DBMS.

However, this approach does not allow the use of linguistic modifiers in the query. This test is interesting since users typically use such linguistic terms and it does not take into account the aggregate queries.

In the approach proposed in [12], clusters generated for each relievable attribute are not stored in the DBMS catalog. Thus, the maintainability of this meta-base is no longer a problem. Indeed, in order to draw the concept lattice, core of FCA, they must simply load an XML file that can retrieve all the information necessary to trace these structures.

However, this approach has limitations in the structures they use. We mainly include:

- The number of concepts generated,
- The response time used to generate approximate answers, and
- The not taking into account aggregate queries.

The approach proposed in [21] allows classical querying (Boolean) on broad comic returning relevant answers in the shortest time for aggregate queries. It aims to gradually give approximate results when the query execution until all data has been processed.

Thus, the user observes the degree of progress of answers and controls the query execution. We are not obliged to wait several minutes for the query result.

The proposed approach combines the advantages of those quoted previously while curing the limits which they present. It contributes several shares in particular:

- The calculation of aggregation for the flexible queries.
- The response time emits with AQ is weaker than that emits without AQ while guaranteeing the answer exactitude.
- The processing of the case of empty answer for a flexible query.
- The no modification of the structure of the DBMS and SQL language.

Table 12 presents a comparative study between the approaches quoted previously and our approach.

Table 12. Comparative Study of Querying Approaches

	Aggregation	Sampling	Flexibility	Exactitude
Query Relaxation			X	
Fuzzification of concepts lattice			X	
Online AQP	X	X		X
Flexible interrogation by AQP	X	X	X	X

5   Conclusion

Aggregation is an increasingly important operation in today's DBMS. As data sets grow larger and both users and user interfaces become more sophisticated, there is a growing emphasis on extracting not just specific data items, but also general characterizations of large subsets of the data. Users want this aggregate information right away, even though producing it may involve accessing and condensing enormous amounts of information.

In traditional querying systems, aggregation is performed in batch mode: a query is submitted, the system processes a large volume of data over a long period of time, and, eventually, the final answer is returned. This archaic approach is frustrating to users and has been abandoned in most other areas of computing.

However, it seems equally clear to us that relational systems can and should be extended to support both online aggregation as well as flexible querying.

In this paper we have proposed an online aggregation in a flexible querying system that permits users to both, specify a flexible query with linguistic terms such as "small", "young'…, observe the progress of their aggregation queries and control execution on the fly.

The type of query is "give the average of the weak salary"

This system makes it possible to turn over quickly approximate answers while holding trying to improve the exactitude of the provided answers.

Our approach includes two steps:

– Pre processing step in which the KB is generated starting from the DB so that it contains membership degrees of each tuple to the relievable attributes.
– Post processing step during which the flexible query is rewritten so that it becomes an AQ. The sampling of the KB consists in extracting some data (tuples). The construction of a sample concepts table is made to release approximate answers and the Error Rate.

For the answer exactitude, we used AQ which support the response time to the detriment of the result exactitude.

In order to improve the response time, we propose in this article to adapt online aggregation technique proposed in [21] whose objective is to gradually give approximate answers while executing the query.

It consists to apply a sampling to the initial data of the DB in order to minimize the disc access and consequently to improve the response time.

Our approach contributes several shares in particular:

- The calculation of aggregation for flexible queries.
- The improvement of response time by guaranteeing the exactitude of the answer.
- The processing of the case of empty answers for a flexible query.
- The no modification of the structure of the DBMS and SQL language.
- The use query execution control.

Compared to the traditional querying systems, the enhancements require changes not only to the user interface of the classic online aggregation, but also to the techniques used for flexible query rewriting, optimization and execution.

We have reported on an initial implementation of online aggregation in a flexible querying system, called FLEXTRA.

This work opens up a number of areas for future works such as:

- A complete implementation of online aggregation in a flexible querying system must be able to handle multi-table (join operator) queries.
- We have spent considerable effort on extending online aggregation past simple joins and to SQL queries with a sub query that is linked to an outer query via an IN, EXISTS, NOT IN, or NOT EXISTS clause.
- The inclusion of some widely used language modifiers like "very" and "approximately" in the query qualification.
- The use of other sampling procedures in order to improve the confidence rate.

References

1. Bosc, P., Pivert, O.: Some Approaches for Relational Databases Flexible Querying. International Journal of Intelligent Information Systems 1, 323–354 (1980)
2. Bosc, P., Liétard, L., Pivert, O.: Bases de Données et Flexibilité: Les Requêtes Graduelles. Techniques et Sciences Informatiques 17(3), 355–378 (1998)
3. Lacroix, M., Lavency, P.: Preferences: Putting More Knowledge Into Queries. In: Proceedings of the 13th VLBD Conference, Brighton, England, pp. 217–225 (September 1987)
4. Chan, C.L.: Decision Support in an Imperfect World. Research Report, 100–102 (1982)
5. Rabitti, F.: Retrieval of Multimedia Documents by Imprecise Query Specification. In: Bancilhon, F., Zhang, J., Thanos, C. (eds.) EDBT 1990. LNCS, vol. 416, pp. 202–218. Springer, Heidelberg (1990)
6. Ichikawa, T., Hirakawa, M.: ARES: A Relational Database with the Capability of Performing Flexible Interpretation of Queries. IEEE Transactions on Software Engineering, 624–634 (1986)
7. Motro, A.: VAGUE: A User Interface to Relational Database That Permits Vague Queries. ACM Transaction Off Information Systems 6(3), 187–214 (1988)
8. Tahani, V.: A conceptual Framework for fuzzy Query Processing: A step Toward Very Intelligent Database Systems. Information Processing and Management 13, 289–303 (1977)
9. Chu, W.W., Yang, H., Chiang, K., Minock, M., Chow, G., Larson, C.: CoBase: A Scalable and Extensible Cooperative Information System. Journal of Intelligence Information Systems 6, 223–253 (1996)

10. Ounelli, H., Belhadjahmed, R.: Interrogation flexible et coopérative d'une BD par abstraction conceptuelle hiérarchique. In: Inforsid, Biarritz France, pp. 41–56 (May 2004)
11. Hachani, N., Ben Hassine, M.-A., Chettaoui, H., Ounelli, H.: Cooperative Answering of Fuzzy Queries. J. Comput. Sci. Technol. 24(4), 675–686 (2009)
12. Grissa-Touzi, A., Sassi, M., Ounelli, H.: An Innovative Contribution to Flexible Query Through the Fusion of Conceptual Clustering, Fuzzy Logic, and Formal Concept Analysis. I. J. Comput. Appl. 16(4), 220–233 (2009)
13. Wille, R.: Restructuring lattice theory: An approach based on hierarchies of concepts. Ordered sets 23, 445–470 (1982)
14. Han, J., Kamber, M.: Data mining: Concepts and Techniques. Morgan Kaufmann Publishers, San Francisco (2001)
15. Chakrabati, S., Garofalakis, M., Rastogi, R., Shim, K.: Approximate query processing using wavelets. Springer, Heidelberg (2001)
16. Ioannidis, Y.E., Poosala, V.: Histogram-Based Approximation of Set-Valued Query-Answers. In: Proceedings of the 15th VLBD Conference, Edinburgh, Scotland, pp. 174–185 (September 1999)
17. Olken, F.: Random sampling from databases, PhD Thesis, University of California, Berkeley (1993)
18. Zadeh, L.A.: Fuzzy sets. Information Control 8, 338–353 (1965)
19. Loney, K.: Oracle Database 10 g: The Complete Reference. Oracle Press (2004)
20. Gammoudi., M.M.: Décomposition conceptuelle des relations binaires et ses applications, Habilitation en Informatique, Faculté des Sciences de Tunis (2005)
21. Haas, P.J., Hellerstein, J.M., Wang, H.J.: Online aggregation. In: Proceeding of ACM-SIGMOD International Conference on Management of Data, Tucson, Arizona (May 1997)
22. Hettich, S., Bay, S.: The UCI KDD archive (1999), http://archive.ics.uci.edu/ml/

Appendix

The domain expert carries out the following application as indicated in Fig. 10:

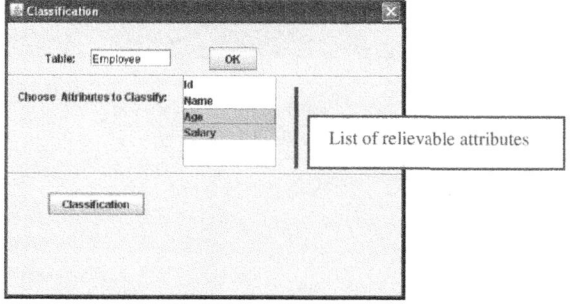

Fig. 10. Classification Wizard

With this wizard, the expert can select his table as well as various relievable attributes to be classified.

Fig. 11 presents the main screen of our DB flexible querying system.

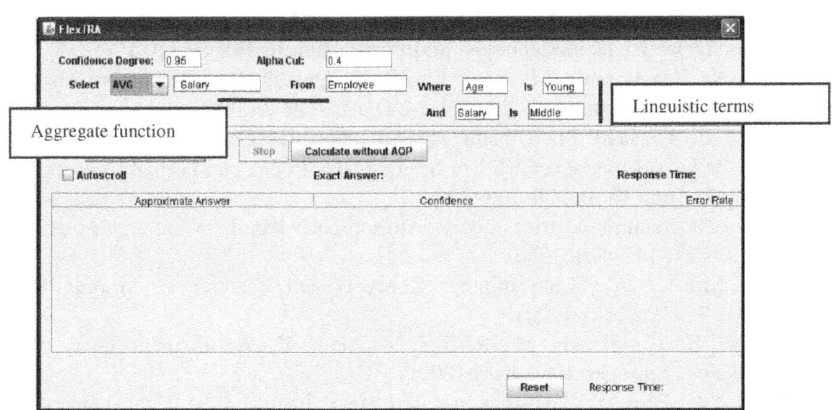

Fig. 11. Main screen of querying system

In this screen, the user provides the following parameters: (see Fig. 11)

- The confidence degree (for example 0.95) which is useful for calculation of the Error Rate or the interval.
- The threshold: the minimum value which is used in order to eliminate the answers which have a small membership degree.
- The aggregation function: "Sum", "Avg", "Count", "Var" etc.
- The attribute to be incorporated,
- The source DB table ;
- The flexible query conditions.

The answers are progressively displayed in the data grid below in the same time of the query execution. In this case, the user is not obliged to await the end of the execution to have the final result. Indeed, after a certain time, the first answer and its Error Rate are displayed and so on, as shown in Fig. 12, until the user stops calculation or which the KB was entirely treated.

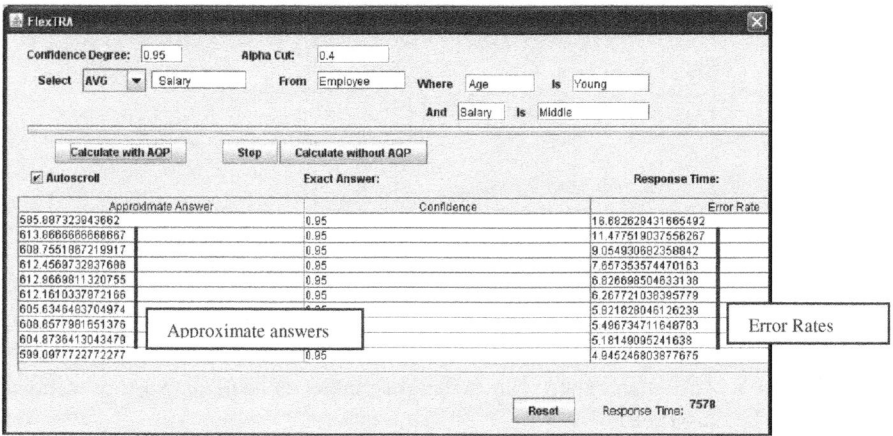

Fig. 12. Display of the approximate answers according to the Error Rate

The answer is composed of two parts:

− Approximate answers.
− Error Rates (the interval in which is the exact answer).
  Example: Answer 1= 585.88 Dt   + /- 16,682   with 95 % Confidence
           Answer 2 = 613.11 Dt +/- 11,477   with 95 % Confidence

If the user wants to know the exact result without using the AQ, he can uses the option "Calculate without AQ" to see the exact result as indicated in Fig. 13.

With the difference of calculation with AQ, classic calculation requires even more time for the answer generation.

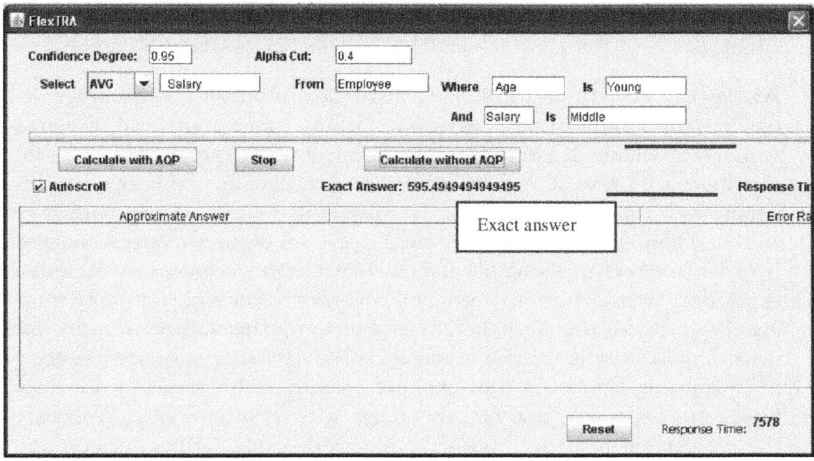

Fig. 13. Display of the exact Answer

Metric-Based Similarity Search
in Unstructured Peer-to-Peer Systems

Akrivi Vlachou[1], Christos Doulkeridis[1], and Yannis Kotidis[2]

[1] Dept. of Computer and Information Science, NTNU
7491 Trondheim, Norway
{vlachou,cdoulk}@idi.ntnu.no
[2] Dept. of Informatics, AUEB
10434 Athens, Greece
kotidis@aueb.gr

Abstract. Peer-to-peer systems constitute a promising solution for deploying novel applications, such as distributed image retrieval. Efficient search over widely distributed multimedia content requires techniques for distributed retrieval based on generic metric distance functions. In this paper, we propose a framework for distributed metric-based similarity search, where each participating peer stores its own data autonomously. In order to establish a scalable and efficient search mechanism, we adopt a super-peer architecture, where super-peers are responsible for query routing. We propose the construction of metric routing indices suitable for distributed similarity search in metric spaces. Furthermore, we present a query routing algorithm that exploits pruning techniques to selectively direct queries to super-peers and peers with relevant data. We study the performance of the proposed framework using both synthetic and real data demonstrate its scalability over a wide range of experimental setups.

1 Introduction

Similarity search in metric spaces has received significant attention in centralized settings [7,19], but also recently in decentralized environments [14,17,26]. A prominent application is distributed search for multimedia content, such as images, video or plain text. For example, the SAPIR project[1] aims at creating an infrastructure for similarity search, by adopting a fully distributed architecture based on peer-to-peer (P2P) technology [2]. In such a distributed search engine, the objective is to find all objects that are similar to a given query object, such as a digital image or a text document. Objects are represented in a high-dimensional feature space and a metric distance function defines the similarity of two objects. In order to provide effective search over multimedia content, techniques for retrieval based on generic metric distance functions are required.

Existing approaches for P2P metric-based similarity search mainly rely on a structured P2P overlay, which is used to intentionally store objects to peers

---

[1] http://www.sapir.eu/

A. Hameurlain et al. (Eds.): TLDKS V, LNCS 7100, pp. 28–48, 2012.

by means of a distributed hash table (DHT) implementation [17,26]. The aim is to achieve high parallelism and share the high processing cost over a set of cooperative computers. In contrast, in this paper, we focus on the scenario of autonomous peers that store multimedia content and collaborate in order to process similarity queries over distributed data. The main differentiating factor is that multimedia content remains stored on the peer that owns it, thus respecting privacy requirements, instead of being stored at arbitrary peers as dictated by the DHT of the underlying structured P2P system.

Our work is motivated by the design of a P2P architecture for image retrieval, as proposed in [33]. In more details, content providers are simple peers that keep multimedia content, which is usually user-generated at the peer. Each peer joins the collaborative search engine, by connecting to one of the information brokers that act as super-peers, using the basic bootstrapping protocol. Super-peers are responsible for establishing the search mechanism. Any peer issues similarity queries for multimedia content (images) and the execution of queries (query routing) is performed at super-peer level, thus directing queries to peers with relevant content.

In this paper, we propose a framework for distributed metric-based similarity search that relies on a super-peer architecture, assuming that cooperative peers store and index their data in an autonomous manner. Each peer must be able to process efficiently similarity queries based on its locally stored data. Thus, each peer indexes its local data by using the M-Tree [9]. The M-Tree consists of a hierarchy of hyper-spheres and is one of the most commonly used centralized indexing techniques for searching in metric spaces. When a peer connects to a super-peer, it publishes the set of hyper-spheres stored at the root of its M-Tree to its super-peer, as a summarization of the stored data. The super-peers store the collected hyper-spheres using an M-Tree index, in order to direct queries only to relevant peers efficiently, thus establishing a *peer selection mechanism*. Capitalizing on their local metric index structures, super-peers exchange summary information to construct metric-based routing indices, which improve the performance of query routing significantly. Then, given a range query, this *super-peer selection mechanism* enables efficient query routing only to that subset of super-peers that are responsible for peers with relevant query results.

The contributions of this paper are the following:

1. We propose a framework for distributed metric-based similarity search, under the assumption that the basic indexing method available on peers and super-peers is the M-Tree index.
2. Our framework relies on the construction of metric-based routing indices for similarity queries in metric spaces, over a super-peer architecture.
3. We propose a routing mechanism that selectively routes similarity queries only to those peers and super-peers that store relevant data.
4. We assess the feasibility of out framework, by means of an experimental evaluation, employing both synthetic and real datasets.

In [32], we shortly described our framework for metric-based similarity search in P2P systems. In this paper, we extend this work substantially by presenting the proposed metric-based routing indices in detail. We also present our novel algorithms for query routing that capitalize on pruning properties to achieve efficiency. Moreover, we discuss maintenance issues, such as updates and peer churn. Finally, we provide a more extensive experimental evaluation.

The rest of this paper is structured as follows: Section 2 provides an overview of related work. In Section 3, the preliminaries are presented, while Section 4 describes the construction of the routing indices based on the exchanged summary information. Then, in Section 5, we present our query routing mechanism. Section 6 deals with maintenance issues. The experimental evaluation is presented in Section 7 and we conclude the paper in Section 8.

2   Related Work

Similarity search in metric spaces has wide applicability in centralized domains. For an overview of relevant algorithms and techniques, see [7,19].

2.1   P2P Metric-Based Similarity Search

Recently, metric similarity search has also attained increased interest in P2P systems [14,17,26,33]. There are two main approaches for facilitating metric similarity search in distributed environments.

The first approach includes systems like MCAN [17] and M-Chord [26], relying on an underlying structured P2P network, namely CAN [28] and Chord [31] respectively. Both techniques focus on parallelism for query execution, motivated by the fact that in real-life applications, a complex distance function is typically expensive to compute. MCAN uses a pivot-based technique that maps data objects to an N-dimensional vector space, while M-Chord uses the principle of iDistance [20] to map objects into one-dimensional values. Data preprocessing – clustering and mapping – is done in a centralized fashion, and only then data is assigned to peers. Relevant to this work, Batko et al. [4] present a comparative experimental evaluation of distributed similarity search techniques. VPT* and GHT* [3] are two distributed metric index structures that are included in the comparison together with MCAN and M-Chord. Later, bulk loading for structured P2P systems has been proposed [13], focusing on peer splits. Recently, the Metric Index (M-Index) [25] has been proposed as a general approach for metric-based data management.

In the second category, peers store data in an autonomous manner and an architecture that supports efficient similarity search using a super-peer architecture was presented in [33]. In SIMPEER [14], P2P metric-based indexing is supported using the iDistance [20] technique. An extension of SIMPEER for recall-based range queries is presented in [15]. In contrast, this paper provides an alternative technique for similarity search in metric spaces, based on a popular metric index (M-Tree) for data access both on peers and super-peers. More importantly, relying on M-trees as routing indices, we present pruning techniques that enhance the performance of query routing.

2.2 Similarity Search in P2P Systems

Apart from the aforementioned research, there exist several approaches for P2P similarity search that do not focus on metric spaces.

In unstructured P2P systems, a general solution for P2P similarity search for vector data is proposed in [1], named SWAM. Peers autonomously store their data, and efficient search is based on an overlay topology that brings nodes with similar content together. However, SWAM is not designed for metric spaces. Content-based similarity search using a hierarchical P2P network is studied in [16]. A P2P framework for multi-dimensional indexing based on a tree-structured overlay is proposed in [22]. LSH forest [5] stores documents in the overlay network using a locality-sensitive hash function to index high-dimensional data for answering approximate similarity queries. Another approach that focuses of semantic content search over distributed document collections is described in [29], where a hierarchical summary index is built over a super-peer architecture. In [12], Datta et al. study range queries over trie-structured overlays.

Most approaches that address range query processing in P2P systems rely on space partitioning and assignment of specific space regions to certain peers. A P2P framework for multi-dimensional indexing based on a tree structured overlay is proposed in [22]. A load-balancing system for range queries that extends Skip Graphs is presented in [30]. The use of Skip Graphs for range query processing has also been proposed in [18]. Several P2P range index structures have been proposed, such as Mercury [6], P-tree [10], BATON [21]. A variant of structured P2P for range queries that aims at exploiting peer heterogeneity is presented in [27]. In [24], the authors propose NR-tree, a P2P adaptation of the R*-tree, for querying spatial data. Routing indices stored at each peer are used for P2P similarity search in [23]. Their approach relies on a freezing technique, i.e. some queries are paused and can be answered by streaming results of other queries. Recently, in [11], P-Ring is proposed as an indexing structure that enables range query processing.

3 Preliminaries

In this section, we provide the necessary preliminaries for our framework. We start with the system architecture, then we present the data model, the query types and a short overview of the M-Tree index. An overview of the symbols used can be found in Table 1.

3.1 System Overview

We assume an unstructured P2P network that consists of N_p peers. Some peers have special roles, due to their enhanced features, such as availability, stability, storage capability and bandwidth capacity. These peers are called super-peers SP_i ($i = 1..N_{sp}$), and they constitute only a small fraction of the peers in the

Table 1. Overview of symbols

Symbols	Description
d	Data dimensionality
n	Dataset cardinality
N_p	Number of peers
N_{sp}	Number of super-peers
DEG_p	Degree of peer
DEG_{sp}	Degree of super-peer
$dist()$	Distance function
$R(q,r)$	Range query
N	Node of M-Tree
p	Representative object of N
$r(p)$	Covering radius of p
T	Reference to child node of N

network, i.e. $N_{sp} << N_p$. Peers that join the network directly connect to one of the super-peers. Each super-peer maintains links to peers, based on the value of its degree parameter DEG_p, which is the number of peers that it is connected to. In addition, a super-peer is connected to a limited set of at most DEG_{sp} other super-peers ($DEG_{sp} < DEG_p$).

3.2   Metric Spaces

In our system, peers that join the network autonomously store their own data. Each peer maintains its own data objects, while the features extracted from the objects are represented as d-dimensional points. Thus, each peer P_i holds n_i d-dimensional points, denoted as a set S_i ($1 \leq i \leq N_p$). Assuming horizontal data distribution to the N_p peers, the size of the complete set of points is $n = \sum_{i=1}^{N_p} n_i$ and the dataset S is the union of all peers' datasets S_i ($S = \cup S_i$). Notice that our techniques are applicable also in the case of peers storing overlapping data, i.e., $S_i \cap S_j \neq \emptyset$.

Definitions and Query Types. Similarity search in metric spaces focuses on supporting queries that retrieve objects similar to a query point, when a metric distance function $dist$ measures the objects' (dis)similarity. More formally, a metric space is a pair $M = (\Delta, dist)$, where Δ is a domain of feature values and $dist$ is a distance function with the following properties:

1. $dist(p, q) > 0$, $q \neq p$ and $dist(p, p) = 0$ (non negativity),
2. $dist(p, q) = dist(q, p)$ (symmetry),
3. $dist(p, q) \leq dist(p, o) + dist(o, q)$ (triangle inequality).

The properties of the metric distance function express that a smaller distance between two objects means higher similarity. Therefore, the distance between identical objects should be zero, otherwise the distance has a positive value.

The similarity of two objects is symmetrical, and thus the distance function must also be symmetrical. The triangle inequality guarantees that the distance between two objects p and q is always smaller than the sum of the distances of p and q to any other object o.

Similarity search in metric spaces involves two different types of queries, namely *range* and *nearest neighbor* queries. Range queries are specified by a query object q and a range value r, and the answer set is defined to contain all the objects o from the dataset that have a distance to the query object q smaller than or equal to r:

Range Query $R(q, r)$: Given a query object q and a radius r, a point $p \in S$ belongs to result set R_q^r of the range query, if $dist(q, p) \leq r$.

A range query $R(q, r)$ can be interpreted as "retrieve all objects that are within distance r to q". The k-nearest neighbor (k-NN) query does not require a user to provide a radius for the query and is therefore easier to express than the similarity range query. The k-nearest neighbor query returns the k most similar data points from the dataset and is defined as follows:

k-nearest Neighbor Query $NN_k(q)$: Given a query object q and a positive integer k, the result set NN_q^k of the k-nearest neighbor is a set, such that $NN_q^k \subseteq S$, $|NN_q^k| = k$ and $\forall u, v : u \in NN_q^k, v \in S - NN_q^k$ it holds that $dist(q, u) \leq dist(q, v)$.

An example of a k-nearest neighbor query is "retrieve the k objects in S which are closest in distance to a given object". Given a query object q, a k-nearest neighbor query is equivalent to a range query specified by query point q and a radius equal to the distance of the k-th nearest neighbor. In this paper, we focus on range queries since k-NN queries can be transformed to range queries, if the distance of the k-th nearest neighbor is known. Radius estimation techniques for distributed nearest neighbor search have been studied in [14].

M-Tree. The M-Tree [9] is a distance-based indexing method, suitable for disk-based implementation. An M-Tree can be seen as a hierarchy of metric regions, also known as hyper-spheres or balls, as depicted in Fig. 1. More precisely, all the objects being indexed are referenced in the leaf nodes, while an entry in a non-leaf node stores a pointer to a node at the next lower level along with summary information about the objects in the subtree being pointed at. The objects in the internal nodes are database objects that are chosen (during the insertion) as representative points. For a non-leaf node N, the entries are quad-tuples $\{(p, r(p)), D, T\}$, where p is an representative object, $r(p)$ is the corresponding covering radius, D is a distance value, and T is a reference to a child node of N. The basic property is that for all objects o in the subtree rooted at T, we have $dist(p, o) \leq r(p)$. For each non-root node N, let object p' be the parent object, i.e. the object in the entry pointing to N. The distance value stored in D

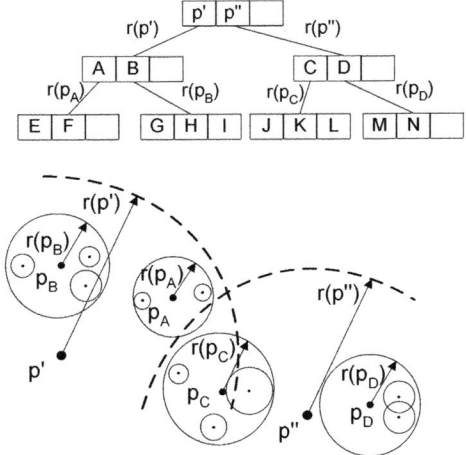

Fig. 1. Example of M-Tree

is the distance $dist(p, p')$ between p and the parent object p' of N. These parent distances allow more efficient pruning during search than would otherwise be possible. Similarly, for a leaf node N, the entries consist of pairs of the form (o, D), where o is a data object and D is the distance between o and the parent object of N.

The M-Tree is built in a bottom-up fashion and heuristics are presented in [9] for choosing the child node to insert an object and for splitting overflowing nodes. Bulk-loading strategies [8] have also been developed for use when an M-Tree must be built for an existing set of data objects. Moreover, the M-Tree is a dynamic index structure, therefore it can efficiently support updates. Range queries $R(q, r)$ for query object q and query radius r can be performed on the M-Tree using a depth-first search algorithm, initiated at the root. Entries of intermediate nodes N are pruned, when the query has no overlap with the ball represented by the the representative object p and its covering radius $r(p)$.

4   Routing Information for P2P Similarity Search

In this section, we first state the objectives that need to be attained by our framework to enable efficient similarity search. Then, we present the routing indices that are built at super-peer level. To elaborate on this, we first describe the local indices maintained at super-peer level, and then we introduce the metric-based routing indices.

4.1   Objectives for P2P Similarity Search

In applications like multimedia retrieval that require efficient similarity search, a server stores a collection of data objects, such as images, which refer to a high-dimensional metric space and a distance function provides a measure of (dis)similarity. In order to support efficient similarity search over the stored data objects, an indexing technique, such as the M-Tree, is required. In our scenario, for distributed image similarity search, we assume that a set of servers, also mentioned as peers, store their data by using an M-Tree. Thus, each server is able to efficiently support similarity queries. The remaining challenge is to support efficient similarity search over the data stored at all peers in a distributed manner.

In this distributed architecture, the individual objectives that need to be attained for efficient processing of similarity queries are: 1) minimizing the number of required messages to retrieve the result, 2) minimizing the number of contacted peers (super-peers) that are involved in processing a similarity query, and 3) minimizing the maximum hop count for a given query, thereby minimizing the associated latency.

The first goal reflects the scalability of the system, in terms of consumed network resources. Obviously, the number of required messages should be kept small, in order to increase the number of queries that can be processed using the available bandwidth. The second goal relates to queries being processed only by peers and super-peers that actually store relevant results. Finally, the last goal is to achieve low latency for queries and minimize the total response time.

In order to achieve the aforementioned objectives, queries have to be forwarded during query processing only to peers and super-peers that may contribute to the query result set, while avoiding to contact peers and super-peers that store data that are not relevant to the query. Thus, instead of flooding queries at super-peer level, we build routing indices that describe the data that is available through each neighboring super-peer. Furthermore, each super-peer maintains information that summarizes the data available at each connected peer. Towards this goal, we use the hyper-spheres of the local M-Trees at each peer.

In the following, we first describe the local indexing at each super-peer (Section 4.2) which creates indices at each super-peers that can be used to determine which peers store relevant data. Then, we proceed by presenting our novel routing indices at each super-peer (Section 4.3) that summarize the data available through each neighboring super-peer.

4.2   Local Indexing

Each peer P_i that connects to a super-peer SP_j publishes a summary of its data, in order to make its content searchable by other peers. In our framework, we take advantage of the existing M-Tree index and each peer P_i publishes to its responsible super-peer SP_j, the hyper-spheres contained in the root of its M-Tree, as a summary of the stored data. This set of hyper-spheres covers all data objects stored at P_i, thus SP_j is able to determine if P_i stores data relevant to a potential range query, by searching for hyper-spheres that overlap with the query.

SP_j needs to support efficient retrieval of peer hyper-spheres, and consequently selection of the peers that store relevant data to a similarity query. For this purpose, SP_j inserts the collected hyper-spheres into a local M-Tree, also mentioned as *super-peer M-Tree*. This enables efficient similarity search over all data stored by peers associated to SP_j by contacting only the peers that store data that may appear in the query result set.

4.3   Routing Indices

The remaining challenge is to construct routing indices for processing similarity queries over the entire super-peer network. For this purpose, each super-peer maintains an M-Tree, also called *routing M-Tree*, to store hyper-spheres (collected from other super-peers) that describe the data accessible through each neighbor in the super-peer topology. The construction of routing indices at super-peer level is achieved in the following way.

A super-peer SP_i sends the descriptions of the hyper-spheres contained in the root of the super-peer M-Tree to its neighbors. This message has the following format: $(msgId, \{(p_m, r(p_m))\})$, where $msgId$ is an identifier that is unique for each SP_i, and $\{(p_m, r(p_m))\}$ represents the set of SP_i's hyper-spheres corresponding to the root of SP_i's M-Tree. Each hyper-sphere is defined by a representative object p_m and the corresponding covering radius $r(p_m)$. Each neighboring super-peer SP_j that receives a set of hyper-spheres for the first time performs two operations. First, SP_j stores locally the hyper-spheres in the routing M-Tree and attaches to them the identifier of the neighboring super-peer SP_i, from which the hyper-spheres were received. Second, SP_j propagates the hyper-spheres to all its neighbors, except for the one it received them from (SP_i). Any super-peer SP_k that is contacted by SP_j performs the same operations. However, notice that SP_k stores in its routing M-Tree the identifier of its neighbor SP_j together with the hyper-spheres, and not the identifier of the owner super-peer SP_i.

This construction protocol works also for network topologies that contain cycles. Since hyper-spheres of any super-peer SP_i are accompanied by a unique $msgId$, each recipient super-peer SP_k can perform duplicate elimination, in case SP_i's hyper-spheres are also received from a different network path. Notice that the granularity of the routing information stored at any super-peer is at the level of its neighbors $O(DEG_{sp})$ and not at the level of the network $O(N_{sp})$. Therefore, the constructed routing indices are scalable with network size.

We now elaborate more on the internal structure of nodes in the routing M-Tree. For internal nodes, the routing M-Tree entry is $\{(p, r(p)), D, T\}$, where $(p, r(p))$ is the representative object p and its covering radius $r(p)$, D is the distance to the parent object and T is a reference to a subtree. For leaf nodes, the routing M-Tree entry is $\{(p, r(p), SP(p)), D\}$, where $(p, r(p), SP(p))$ consists of the representative object p, its covering radius $r(p)$, and $SP(p)$ is the neighbor super-peer responsible for the hyper-sphere, whereas D is the distance to the parent object.

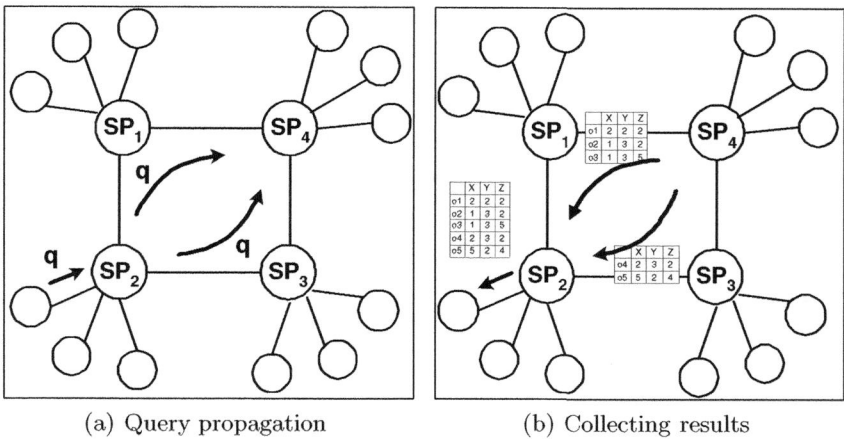

(a) Query propagation   (b) Collecting results

Fig. 2. Super-peer query processing

5   Distributed Similarity Search

Our framework creates routing M-Trees on each super-peer and supports efficient query processing, in terms of local computation costs, communication costs
and overall response time. In this section, we present our novel query routing
algorithm that exploits the proposed routing indices.

A query may be posed by any peer P_q and is propagated to the associated
super-peer SP_q, which becomes responsible for local query processing and query
routing, and finally returns the result set to P_q. Each super-peer SP_i that receives the query performs two tasks: 1) SP_i routes the query to a subset of
its neighboring super-peers, and 2) SP_i processes the query locally, in order to
retrieve results from its local peers.

Afterwards, the relevant data is collected at SP_i and sent back to the neighboring super-peer from which the query was received. Finally, SP_q collects all
results of its neighboring super-peers and sends the result set back to the peer
P_q that posed the query. A high-level view of the query processing functionality
is depicted in Fig. 2. The query propagation is shown in Fig. 2(a), starting from
P_q and SP_q (in the example SP_2) to the rest of the super-peer network. Fig. 2(b)
depicts the collection of query results back to SP_2.

We first elaborate on the details of local processing, and then the query routing
mechanism is presented.

5.1   Super-Peer Local Query Processing

Given a range query $R(q, r)$, query processing at SP_i is performed by exploiting
the summary information stored in the super-peer M-Tree. The aim is to retrieve
the subset of local peers that need to be contacted. The peers that store data
enclosed in the range query $R(q, r)$ have to be contacted, since these results are

necessary to be retrieved and reported back to SP_q, in order to form the exact and complete result set. Therefore, SP_i uses its super-peer M-Tree to identify hyper-spheres of peers that intersect with the query. Recall that the retrieved hyper-spheres contain the peer identifier of the owner peer. Thus, the subset of peers that can contribute to the query result is determined and the range query is forwarded to the corresponding peers. This enables efficient similarity search over all data stored by peers associated to SP_i, since the query is posed only to peers having data that may appear in the result set, essentially forming an effective *peer selection mechanism* at super-peer SP_i. Each recipient peer processes the query using its local M-Tree, in the traditional way of processing range queries in M-Trees. Consequently, each peer reports its results to SP_i, which in turn is responsible for returning the results to SP_q.

5.2   Query Routing

After having described the local query processing on each super-peer, we proceed to present the details on query routing at super-peer level. Henceforth, we assume that each super-peer that receives the query also performs local query processing, as described above, therefore we omit the details of local query processing from the following discussion.

Given a range query $R(q, r)$, the querying super-peer SP_q needs to selectively propagate the query to a fraction of its neighboring super-peers as will be described shortly. Each intermediate super-peer SP_i that receives the query repeats the same process. The routing algorithm on any super-peer SP_i is based on its routing M-Tree. During query routing the summary information that is stored at the routing M-Tree of each super-peer SP_i is exploited. The aim of the routing algorithm is to retrieve all hyper-spheres that intersect with the query and the neighboring super-peers that should be contacted respectively. As a consequence, when a super-peer SP_r receives a range query $R(q, r)$, SP_r uses the routing M-Tree to efficiently retrieve all hyper-spheres that have an overlap with the query. Then, the set of neighbor super-peers is determined and the query is forwarded to them only. This forms the *super-peer selection mechanism* that enables routing of queries at super-peer level.

We now describe in more details how the set of relevant neighboring super-peers are retrieved from the M-Tree. Given a range query $R(q, r)$ defined by query object q and radius r, query routing is performed on the routing M-Tree by means of a depth-first traversal, initiated at the root. Let N be a node that is being visited with entry $\{(p, r(p)), D, T\}$, and let p' be its parent object. D represents the distance of p to its parent p', and T is a reference to a child node. The following two observations can be used for discarding subtrees during the M-Tree traversal. The first observation exploits the pre-computed distances $D = dist(p, p')$ to parent objects, in order to avoid computing the actual distance $d(p, q)$.

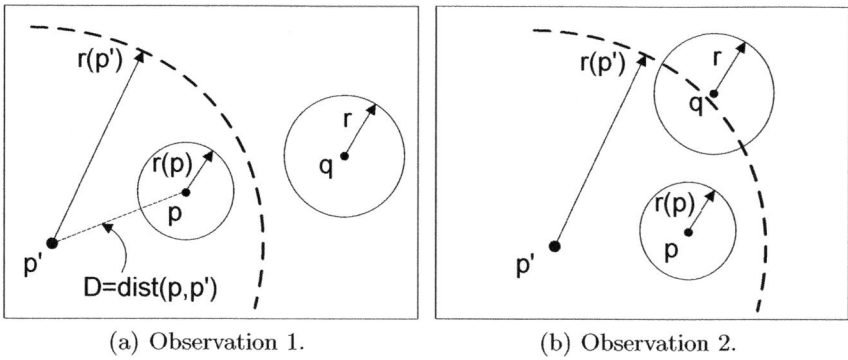

(a) Observation 1.   (b) Observation 2.

Fig. 3. Conditions for pruning a subtree of routing M-Tree

Observation 1. Given a range query $R(q, r)$, if $|dist(p', q) - dist(p, p')| > r + r(p)$, then the subtree pointed by T can be safely pruned from the search.

The interpretation of the first observation is based on the fact that $|dist(p', q) - dist(p, p')| - r(p)$ is a lower bound of the distance of any object in the subtree pointed at by T. Thus, if the lower bound is greater than r, then no object in this subtree can be in the range (Fig. 3(a)). If the condition of Observation 1 is not satisfied, the distance $dist(p, q)$ must be computed. However, after having computed $dist(p, q)$, we can still avoid visiting the node pointed at by T, if the lower bound on the distance from q to any object p in T is greater than r. This is the case if $dist(p, q) > r + r(p)$, since the lower bound is $dist(p, q) - r(p)$ (Fig. 3(b)).

Observation 2. Given a range query $R(q, r)$, if $dist(p, q) > r + r(p)$, then the subtree pointed by T can be safely pruned from the search.

Capitalizing on these observations, Algorithm 1 describes the query routing process at a super-peer, in terms of determining the neighbor super-peers that need to be queried. It takes as input the range query $R(q, r)$ and the current tree node N. For each object p that belongs to node N (line 4), Observation 1 is used to prune the subtree T pointed by p (line 5). If T cannot be pruned, the distance $dist(p, q)$ is computed (line 6) and then the condition of Observation 2 is checked (line 7), in order to discard the subtree T pointed by p. If this condition does not hold, then we distinguish between two cases (line 8). If N is not a leaf node, then the algorithm is invoked on the subtree T (line 9). Otherwise, if N is a leaf node, then $SP(p)$ is added to the result (line 11). At the end of the algorithm, the list of neighboring super-peers to which the query should be routed is determined.

Algorithm 1. Query Routing QR

1: **Input:** N:node, q:query point, r:search radius
2: $RES = \emptyset$
3: let p' be the parent object of node N
4: **for** $(\forall p \in N)$ **do**
5:     **if** $(|dist(p',q) - dist(p,p')| \leq r + r(p))$ **then**
6:        compute $dist(p,q)$
7:        **if** $(dist(p,q) \leq r + r(p))$ **then**
8:           **if** $(N$ is not a leaf$)$ **then**
9:             $QR(T,q,r)$
10:          **else**
11:             $RES = RES \cup \{SP(p)\}$
12:          **end if**
13:        **end if**
14:     **end if**
15: **end for**
16: **return** RES

6   Maintenance

In a dynamic P2P environment, maintenance of routing information is important, especially in the presence of data updates, peer joins and failures. In this section, we first discuss the maintenance cost of the routing indices. Afterwards, we describe how churn affects the proposed framework.

6.1   Data Updates

Maintenance of routing indices is triggered by data insertion, updates and deletions that occur at any peer. Any change of the data stored at a peer P_i causes updates to P_i's local M-Tree. However, as long as the peer's hyper-spheres contained in the root of the M-Tree do not change, such updates do not need to be propagated to any other super-peer.

In the case that a root hyper-sphere description changes, then P_i's responsible super-peer has to be informed. The super-peer updates its local M-Tree index and checks if the update changes the respective hyper-sphere descriptions at root level. Only if such a change occurs, need the other super-peers be updated. Otherwise, the change only affects the particular super-peer. Notice that if hyper-spheres shrink, then even if the super-peers are not updated immediately, the framework still provides correct answers to queries, at the cost of contacting more super-peers. In practice, a super-peer informs its neighbors about changes by broadcasting the modification in a similar way to the construction phase.

To summarize, data updates incur maintenance costs only if the hyper-spheres of a peer root, and eventually the super-peer's root hyper-sphere, are modified.

6.2   Churn

The existence of super-peers makes the system more resilient to failures compared to other P2P systems. Super-peers have stable roles, but in the rare case that a super-peer fails, its peers can detect this event and connect to another super-peer using the basic bootstrapping protocol.

On the other hand, a peer failure may cause the responsible super-peer to update its hyper-spheres. Only if churn rate is high, these changes need to be propagated to other super-peers. Even if updates are not propagated immediately after a peer fails, the only impact to our system is that the cost of searching is increased (i.e. super-peers no longer holding relevant results may be contacted), but the validity of the result is not compromised.

As already mentioned, a peer joins the network by contacting a super-peer using the bootstrapping protocol. The bootstrapping super-peer SP_B uses its routing clusters to find the most relevant super-peer to the joining peer. This is equivalent to the way similarity search is performed over the super-peer network. When the most relevant super-peer SP_r is discovered, the new peer joins SP_r. An interesting property of our approach is that joining peers become members of relevant super-peers, so it is expected as new peers join the system, that clustered datasets are gradually formed, with respect to the assigned super-peers.

7   Experimental Evaluation

In order to evaluate the performance of our approach, we implemented a simulator prototype in Java. The simulations run on 3.8GHz Dual Core AMD processors with 2GB RAM. In order to be able to test the algorithms with realistic network sizes, we ran multiple instances of the peers on the same machine and simulated the network interconnection.

The P2P network topology used in the experiments consists of N_{sp} interconnected super-peers in a random graph topology. We used the GT-ITM topology generator[2] to create well-connected random graphs of N_{sp} peers with a user-specified average connectivity (DEG_{sp}). In our experiments we vary the network size (N_p) from 4000 to 20000 peers, while the number of super-peers varies from 200 to 1000. We also tested different DEG_{sp} values ranging from 4 to 7 and different number of peers per super-peer ($DEG_p = 20 - 100$). In addition, the query selectivity (Q_{sel}) of range queries is varied leading to queries that retrieve between 50 and 200 objects.

We used synthetic data collections, in order to study the scalability of our approach. Both uniform and clustered datasets are employed that were horizontally partitioned evenly among the peers. The uniform dataset includes random points in $[0, 10000]^d$. For the clustered dataset, each super-peer picks randomly a d-dimensional point and all associated peers obtain k_p cluster centroids that follow a Gaussian distribution on each axis with variance 0.05.

---

[2] Available at: http://www.cc.gatech.edu/projects/gtitm/

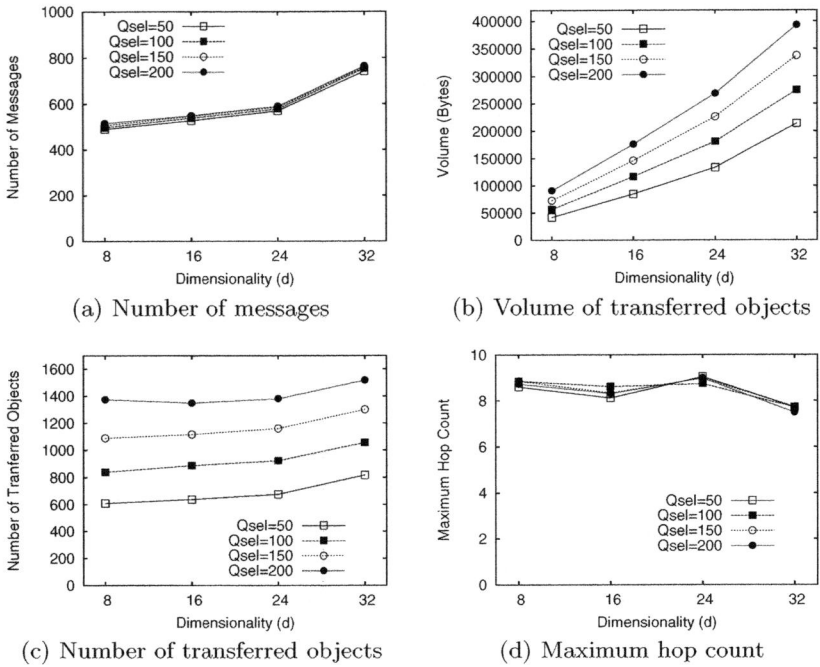

Fig. 4. Scalability with dimensionality for clustered dataset

Thereafter, the peers' objects are generated by following a Gaussian distribution on each axis with variance 0.025, and a mean equal to the corresponding coordinate of the centroid. We conduct experiments varying the dimensionality d (8-32d) and the cardinality n (4M-20M) of the dataset. We keep n/N_p=1000 in all setups. Additionally, we employed a real data collection (VEC), which consists of 1M 45-dimensional vectors of color image features, which was distributed to peers uniformly at random. In all cases, we generate 100 queries uniformly distributed and we show the average values. For each query a peer initiator is randomly selected. Although different metric distance functions can be supported, in this set of experiments we used the Euclidean distance function. We measure: (i) number of messages, (ii) volume of transferred data, (iii) number of transferred objects, (iv) maximum hop count, (v) number of contacted peers, (vi) number of contacted super-peers, and (vii) response time.

7.1 Scalability with Dimensionality

Initially, we focus on the case of clustered dataset. We use a default setup of: N_{sp}=200, N_p=4000, DEG_{sp}=4, n=4M, and the selectivity of range queries ranges from 50 to 200 objects. We study the effect of increasing dimensionality d to our approach. In Fig. 4(a), the number of messages required for searching increases when the dimensionality increases. The volume and the number of

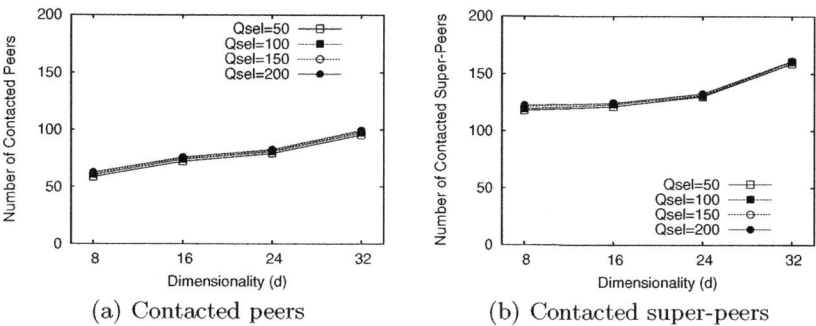

Fig. 5. Contacted peers and super-peers for clustered dataset

transferred objects are depicted in Fig. 4(b) and Fig. 4(c) respectively. Although the number of transferred objects is not significantly affected, the volume increases with dimensionality, as more bytes are necessary for representing objects of higher dimensionality. Notice that the transferred volume remains relatively low, between 100KB and 400KB. In Fig. 4(d), the maximum hop count is depicted, which relates to the latency of the approach. The number of required hops is between 8 and 9, irrespective of the increased dimensionality, which implies that latency is low, considering that the network consists of 4000 peers. Then, in Figs. 5(a) and 5(b), we measure the number of contacted peers and super-peers respectively. Although the number of super-peers that process the query is between 120 and 150, the number of peers is much lower, ranging from 60 to 100 peers.

7.2   Scalability with Network Parameters

In the following, we study the scalability of our approach with respect to the network parameters, by fixing $d=8$. For this purpose, we increase the number of super-peers N_{sp} (Fig. 6(a)), peers N_p (Fig. 6(b)) and the average number of connections per super-peer DEG_{sp} (Fig. 6(c)). We observe that the maximum hop count increases only slightly, always remaining below 12, when the number of super-peers is increased by a factor of 5. On the other hand, in Fig. 6(b), the increasing number of peers only affects the number of contacted peers, however the increase is only marginal compared to the network size. Lastly, increasing the density of the super-peer network ($DEG_{sp}=4$-7) causes more messages to be transferred (500-700), in order to retrieve the result, but again the increase is small even in the case that the network becomes dense ($DEG_{sp}=7$).

7.3   Evaluation for Uniform Data

In Fig. 7, we examine the case of uniform data. Clearly, this is a hard case for our approach, as a query may in worst case have to contact all peers, in order to

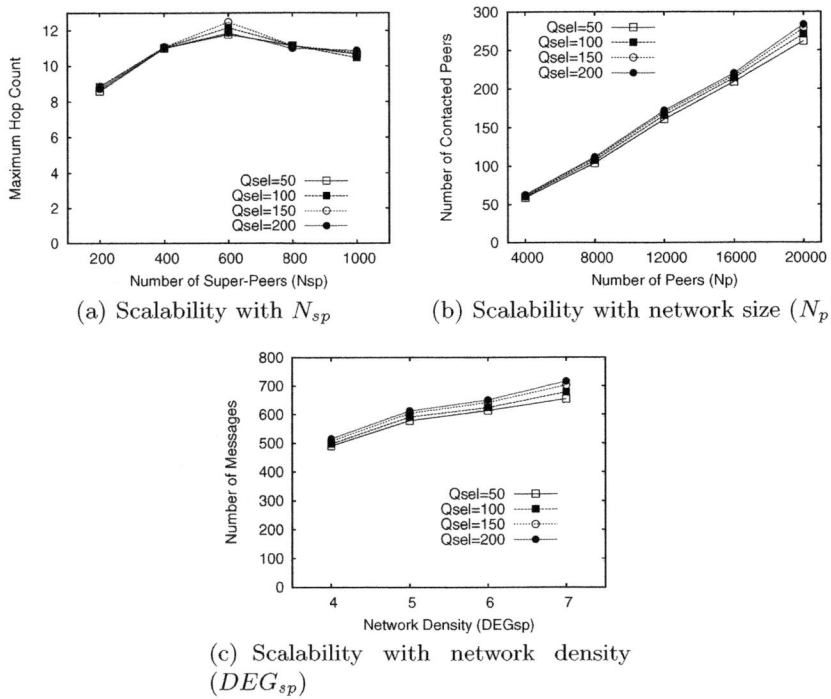

(a) Scalability with N_{sp}

(b) Scalability with network size (N_p)

(c) Scalability with network density (DEG_{sp})

Fig. 6. Scalability for clustered dataset with respect to network parameters

retrieve the correct results. This actually occurs in our experiments, causing also a large number of messages to be sent. However, notice that the maximum hop count, depicted in Figure 7(a), is small (around 6), even smaller than in the case of clustered dataset, since the probability of finding the results in smaller distance increases. We show the volume of transferred data in Fig. 7(b). Compared to the case of the clustered dataset, the total volume transferred increases by a factor of 3-4.

7.4   Evaluation for Real Data

In addition, we evaluate our approach using a real dataset (VEC), which consists of 1M 45-dimensional vectors of color image features. We used a network of 200 super-peers and 1000 peers, thus each peer stores 1000 data points. In Fig. 7(c), the number of contacted peers and super-peers are depicted for increasing query selectivity from 50 to 200 points. The results are comparable to the case of the clustered dataset, but slightly worse, as the VEC dataset is not clustered. However, notice that the absolute numbers are comparable to the results obtained using the synthetic dataset, which is a strong argument in favor of the feasibility of our approach.

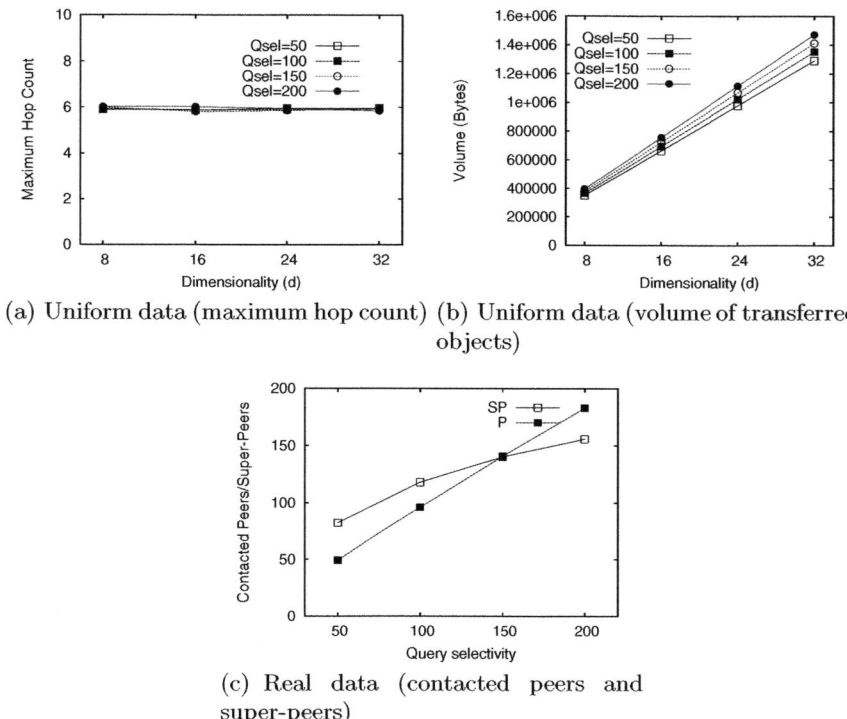

(a) Uniform data (maximum hop count) (b) Uniform data (volume of transferred objects)

(c) Real data (contacted peers and super-peers)

Fig. 7. Experiments with uniform and real datasets

7.5   Maintenance

We performed a series of experiments, in order to study the effect of updates on our system. For this purpose, we updated up to 100% of each peer's data, by deletions of points followed by insertions of new data points that follow the data distribution. Then, we measured the percentage of updates that actually triggered an update of a) the peer's hyper-spheres in the root of the peer M-Tree, and b) the super-peer's hyper-spheres in the root of the super-peer M-Tree. The first is a metric of the update costs from any peer to its super-peer, whereas the second indicates the update cost at super-peer level. The results show that only fewer than 2% of the updates lead to updates from peers to super-peers. More importantly, the percentage of updates that cause an update at super-peer level is only 0.1%. This shows that only a small percentage of data updates have an effect on the super-peer network, and such updates can still be efficiently managed using the protocols described in Section 6.

7.6   Comparison to SIMPEER

In Fig. 8, we study the comparative performance of the proposed framework to SIMPEER [14]. SIMPEER relies on the iDistance indexing technique instead of

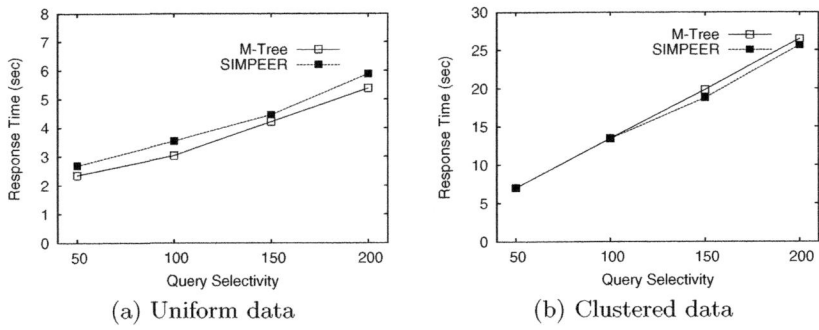

Fig. 8. Comparison to SIMPEER in terms of response time

the M-Tree. We performed a set of experiments using both approaches, assuming a modest 4KB/sec as network transfer rate, and we discuss our main findings here. We employ both a uniform and a clustered dataset, in order to explore the performance of both approaches. In the case of the uniform dataset, our framework outperforms SIMPEER in terms of response time, as depicted in Fig. 8(a). In contrast, when a clustered dataset is used, SIMPEER is marginally better than our framework, as shown in Fig. 8(b). The key factor that determines the individual performance achieved is the routing ability of the set of clusters that iDistance uses to summarize the data, compared to the hyper-spheres of the root of the M-Tree. For the clustered dataset, SIMPEER is able to accurately discover the underlying clusters in the data, resulting in better performance. When the data distribution is uniform, our framework based on M-Trees is more efficient than SIMPEER, since the performance of our metric-based routing indices is not influenced by the absence of a clustering structure in the data.

The M-Tree approach builds the index in a bottom-up manner and the insertion method as well as the block size influences the quality (and the number of) hyper-spheres of the root. On the other hand, iDistance relies on clustering and the employed clustering method influences its overall performance. A generic clustering algorithm, such as k-means, may lead to a poor performance, while an application-specific clustering method may improve the performance of routing. The advantage of our framework is that it does not require the existence of a clustering structure in the data.

8   Conclusions

Similarity search in metric spaces has several applications, such as image retrieval. In such applications that require similarity search in metric spaces, usually a server indexes its data with a state-of-the-art centralized metric indexing technique, such as the M-Tree. In this paper, we study the challenging problem of supporting efficient similarity queries over distributed data in a P2P system. We assume that each peer autonomously maintains its own data indexed by an M-Tree. We propose a framework for distributed similarity search that exploits the local M-Trees, by using the hyper-spheres stored at the M-Tree roots as a summarization of the data. Based on this information, an efficient routing mechanism for similarity queries is built. The experimental results show that our approach performs efficiently in all cases, while the performance of our framework scales with all network and dataset parameters.

References

1. Banaei-Kashani, F., Shahabi, C.: SWAM: a family of access methods for similarity-search in peer-to-peer data networks. In: Proceedings of CIKM 2004, pp. 304–313 (2004)
2. Batko, M., Falchi, F., Lucchese, C., Novak, D., Perego, R., Rabitti, F., Sedmidubský, J., Zezula, P.: Building a web-scale image similarity search system. Multimedia Tools Appl. 47(3), 599–629 (2010)
3. Batko, M., Gennaro, C., Zezula, P.: A Scalable Nearest Neighbor Search in P2P Systems. In: Ng, W.S., Ooi, B.-C., Ouksel, A.M., Sartori, C. (eds.) DBISP2P 2004. LNCS, vol. 3367, pp. 79–92. Springer, Heidelberg (2005)
4. Batko, M., Novak, D., Falchi, F., Zezula, P.: On scalability of the similarity search in the world of peers. In: Proceedings of International Conference on Scalable Information Systems (InfoScale), vol. 20 (2006)
5. Bawa, M., Condie, T., Ganesan, P.: LSH forest: self-tuning indexes for similarity search. In: Proceedings of WWW 2005, pp. 651–660 (2005)
6. Bharambe, A.R., Agrawal, M., Seshan, S.: Mercury: supporting scalable multi-attribute range queries. In: Proceedings of SIGCOMM 2004, pp. 353–366 (2004)
7. Chavez, E., Navarro, G., Baeza-Yates, R., Marroquin, J.L.: Searching in metric spaces. ACM Computing Surveys (CSUR) 33(3), 273–321 (2001)
8. Ciaccia, P., Patella, M.: Bulk loading the M-tree. In: Proceedings of Australasian Database Conference (ADC), pp. 15–26 (1998)
9. Ciaccia, P., Patella, M., Zezula, P.: M-tree: An efficient access method for similarity search in metric spaces. In: Proceedings of the International Conference on Very Large Data Bases (VLDB), pp. 426–435 (1997)
10. Crainiceanu, A., Linga, P., Gehrke, J., Shanmugasundaram, J.: P-tree: a P2P index for resource discovery applications. In: Proceedings of WWW 2004 (2004)
11. Crainiceanu, A., Linga, P., Machanavajjhala, A., Gehrke, J., Shanmugasundaram, J.: P-ring: An efficient and robust p2p range index structure. In: Proceedings of SIGMOD, pp. 223–234 (2007)
12. Datta, A., Hauswirth, M., John, R., Schmidt, R., Aberer, K.: Range queries in trie-structured overlays. In: Proceedings of P2P 2005, pp. 57–66 (2005)
13. Dohnal, V., Sedmidubsky, J., Zezula, P., Novak, D.: Similarity searching: Towards bulk-loading peer-to-peer networks. In: Proceedings of International Workshop on Similarity Search and Applications (SISAP), pp. 87–94 (2008)
14. Doulkeridis, C., Vlachou, A., Kotidis, Y., Vazirgiannis, M.: Peer-to-peer similarity search in metric spaces. In: Proceedings of the International Conference on Very Large Data Bases (VLDB), pp. 986–997 (2007)
15. Doulkeridis, C., Vlachou, A., Kotidis, Y., Vazirgiannis, M.: Efficient range query processing in metric spaces over highly distributed data. Distributed and Parallel Databases 26(2-3), 155–180 (2009)

16. Doulkeridis, C., Vlachou, A., Nørvåg, K., Kotidis, Y., Vazirgiannis, M.: Efficient search based on content similarity over self-organizing p2p networks. Peer-to-Peer Networking and Applications 3(1), 67–79 (2010)
17. Falchi, F., Gennaro, C., Zezula, P.: A Content–Addressable Network for Similarity Search in Metric Spaces. In: Moro, G., et al. (eds.) DBISP2P 2005 and DBISP2P 2006. LNCS, vol. 4125, pp. 98–110. Springer, Heidelberg (2007)
18. Ganesan, P., Bawa, M., Garcia-Molina, H.: Online balancing of range-partitioned data with applications to peer-to-peer systems. In: Proceedings of VLDB 2004, pp. 444–455 (2004)
19. Hjaltason, G.R., Samet, H.: Index-driven similarity search in metric spaces. ACM Transactions on Database Systems (TODS) 28(4), 517–580 (2003)
20. Jagadish, H.V., Ooi, B.C., Tan, K.-L., Yu, C., Zhang, R.: iDistance: An adaptive B^+-tree based indexing method for nearest neighbor search. ACM Transactions on Database Systems (TODS) 30(2), 364–397 (2005)
21. Jagadish, H.V., Ooi, B.C., Vu, Q.H.: Baton: a balanced tree structure for peer-to-peer networks. In: Proceedings of VLDB 2005, pp. 661–672 (2005)
22. Jagadish, H.V., Ooi, B.C., Vu, Q.H., Zhang, R., Zhou, A.: VBI-tree: A peer-to-peer framework for supporting multi-dimensional indexing schemes. In: Proceedings of ICDE 2006, vol. 34 (2006)
23. Kalnis, P., Ng, W.S., Ooi, B.C., Tan, K.-L.: Answering similarity queries in peer-to-peer networks. Inf. Syst. 31(1), 57–72 (2006)
24. Liu, B., Lee, W.-C., Lee, D.L.: Supporting complex multi-dimensional queries in P2P systems. In: Proceedings of ICDCS 2005, pp. 155–164 (2005)
25. Novak, D., Batko, M., Zezula, P.: Large-scale similarity data management with distributed metric index. In: Information Processing and Management (2011)
26. Novak, D., Zezula, P.: M-Chord: a scalable distributed similarity search structure. In: Proceedings of International Conference on Scalable Information Systems (InfoScale), vol. 19 (2006)
27. Ntarmos, N., Pitoura, T., Triantafillou, P.: Range Query Optimization Leveraging Peer Heterogeneity in DHT Data Networks. In: Moro, G., Bergamaschi, S., Joseph, S., Morin, J.-H., Ouksel, A.M. (eds.) DBISP2P 2005 and DBISP2P 2006. LNCS, vol. 4125, pp. 111–122. Springer, Heidelberg (2007)
28. Ratnasamy, S., Francis, P., Handley, M., Karp, R., Schenker, S.: A scalable content-addressable network. In: Proceedings of Conference on Applications, Technologies, Architectures, and Protocols for Computer Communication (SIGCOMM), pp. 161–172 (2001)
29. Shen, H.T., Shu, Y., Yu, B.: Efficient semantic-based content search in P2P network. IEEE Trans. Knowl. Data Eng. 16(7), 813–826 (2004)
30. Shu, Y., Ooi, B.C., Tan, K.-L., Zhou, A.: Supporting multi-dimensional range queries in peer-to-peer systems. In: Proceedings of P2P 2005, pp. 173–180 (2005)
31. Stoica, I., Morris, R., Karger, D., Kaashoek, M.F., Balakrishnan, H.: Chord: A scalable peer-to-peer lookup service for internet applications. In: Proceedings of Conference on Applications, Technologies, Architectures, and Protocols for Computer Communication (SIGCOMM), pp. 149–160 (2001)
32. Vlachou, A., Doulkeridis, C., Kotidis, Y.: Peer-to-Peer Similarity Search Based on M-Tree Indexing. In: Kitagawa, H., Ishikawa, Y., Li, Q., Watanabe, C. (eds.) DASFAA 2010. LNCS, vol. 5982, pp. 269–275. Springer, Heidelberg (2010)
33. Vlachou, A., Doulkeridis, C., Mavroeidis, D., Vazirgiannis, M.: Designing a Peer-to-Peer Architecture for Distributed Image Retrieval. In: Boujemaa, N., Detyniecki, M., Nürnberger, A. (eds.) AMR 2007. LNCS, vol. 4918, pp. 182–195. Springer, Heidelberg (2008)

Adaptive Parallelization of Queries
to Data Providing Web Service Operations

Manivasakan Sabesan and Tore Risch

Department of Information Technology, Uppsala University, Sweden
{msabesan,Tore.Risch}@it.uu.se

Abstract. A data providing web service operation returns a collection of objects for given parameters without any side effects. The Web Service MEDiator (WSMED) system automatically provides relational views of any data providing web service operations by reading their WSDL documents. These views are queried with SQL. In an execution plan, a call to a data providing web service operation may depend on the results from other web service operation calls. In other cases, different web service calls are independent of each other and can be called in any order. To parallelize and speed up both dependent and independent web service operation calls, WSMED has been extended with the adaptive operator *PAP*. It incrementally parallelizes calls to web service operations until no significant performance improvement is measured. The performance of *PAP* is evaluated using publicly available web services. The operator substantially improves the query performance without knowing the cost of calling a web service operation and without extensive memory usage.

Keywords: Adaptive parallelization, Query optimization.

1   Introduction

We define *data providing web service operations* as operations where data are retrieved from a server without side effects for given parameters and returned by the operation as a collection. The Web Service MEDiator (WSMED) [11][12] system provides general query capabilities over any data providing web service operations without any further programming. WSMED automatically generates relational views of web service operations by reading the WSDL documents. These views can be queried and joined with SQL. A web service operation is considered as a high latency function call where the result is a data collection. For a given SQL query, WSMED first generates an initial execution plan calling web service operations. Then, at run time, the initial execution plan is adaptively parallelized in order to speed-up the calls to the operations.

As an example, consider a query to find all the information of the places in some of the US states along with their zip codes and weather forecasts. Four different data providing web service operations can be used for answering this query. First the *GetAllStates* operation from the web service *GeoPlaces* [3] is called to retrieve the states. The *GetInfoByState* operation by *USZip* [15] returns all the zipcodes of a given US State. The *GetPlacesInside* operation by *Zipcodes* [4] retrieves the places located

A. Hameurlain et al. (Eds.): TLDKS V, LNCS 7100, pp. 49–69, 2012.
© Springer-Verlag Berlin Heidelberg 2012

within a given zipcode. Finally, the *GetCityForecastByZip* operation by *CYDNE* [6] returns weather forecast information for a given zip code.

Two operation calls are *dependent* if one of them requires as input, an output from the other one, otherwise they are *independent*. In the above example, the web service operations *GetPlacesInside* and *GetCityForecastByZip* are dependent on *GetInfoByState* but independent of each other, while *GetInfoByState* is dependent on *GetAllStates*. A challenge is to develop methods to speed up queries requiring both dependent and independent web service calls, as in the example. In general such speed-ups depend on some web service properties. Those properties are normally not available and furthermore depend on the network and runtime environments when and where the queries are executed. In such scenarios it is very difficult to base execution strategies on a static cost model, as is done in relational databases.

To improve the response time without a cost model, WSMED automatically parallelizes the web service calls at run time while keeping the data dependencies between them. For each web service operation call the optimizer generates a *plan function* which encapsulates the operation call and makes data transformations such as nesting, flattening, filtering, data conversions, and calls to other plan functions.

For adaptive parallelization of queries with web service operation calls, the algebra operator, *PAP* (Parameterized Adaptive Parallelization) has been added to WSMED. It takes as arguments a collection of independent plan functions along with a stream of parameter values to be processed by the plan functions. For each incoming parameter tuple, PAP starts one process per plan function and then extracts from the incoming tuple the arguments for each of the plan function and sends the extracted argument tuple to the corresponding process. The results from the processes are collected asynchronously and delivered as a stream. The result tuples from *PAP* are formed by combining result tuples from each child process. When a child process has delivered all result tuples in a call it is terminated and another child plan function call is started asynchronously in a new process.

A set of independent plan function calls is processed by a single call to *PAP* taking the collection of independent plan functions as arguments, while dependent plan function calls are processed as sequences of parallelized *PAP* calls.

The *PAP* operator provides adaptation of a distributed execution plan without any central control or cost model. *PAP* dynamically modifies a parallel plan by monitoring the performance of each of its plan function calls. Based on the monitoring new child processes are started until no significant performance improvement is measured. Sequences of *PAP* calls will start sub plans executed in sequence, each of them are locally adapted as well.

In summary the contributions of our work are:

— For a given SQL query, the query processor automatically generates a parallel execution plan calling *PAP* that adaptively parallelizes all kinds of web service operation calls. *PAP* is shown to substantially improve the query performance without any cost knowledge or extensive memory usage compared to other strategies such as caching web services calls or adaptive parallelization strategies restricted to sequences of adaptive processes [11][12].

The rest of this paper is organized as follows. In Section 2, we provide a motivating scenario used in experiments in terms of real web services. Section 3 shows how the query plans are generated. In Section 4 adaptive parallelization using PAP and experimental results are presented. Related work is analyzed in Section 5, and finally Section 6 summarizes and indicates future directions.

2   Motivating Scenario

The class of queries we considered are Select-Project-Join (SPJ) queries calling several web service operations, wso_i. If the kth input argument $a_{i,k}$ for operation wso_i is requires the mth result $r_{l,m}$ of wso_l, wso_i is dependent of wso_l. Otherwise the web service operations are independent. Dependent web service calls may be represented as an acyclic dependency graph (ADG) in which there is a node for each web service operation call and where directed edges represent the dependency between web service operations.

$$\text{input} \begin{cases} \rightarrow \text{WSO}_1 \\ \rightarrow \text{WSO}_2 \rightarrow \text{WSO}_3 \end{cases} \quad \text{- - -}$$

Fig. 1. ADG

For example, in Fig. 1 the ADG illustrates that web service operation wso_3 depends on wso_2, while wso_1 and wso_2 are independent. In WSMED each web service operation is encapsulated by an automatically generated SQL view. To provide relational views of web service operations returning complex objects, such an automatically generated SQL view flattens the result from a web service call. Some of the attributes of these SQL views are input parameters. In the generated views web service operation argument values being complex object are first constructed before the operation is called.

The PAP operator is evaluated for the class of queries illustrated by Fig. 1, where there are both dependent and independent web service operation calls. We made experiments with two particular queries of this class that calls different web service operations provided publicly available service providers.

2.1   Query1

The example SQL *Query1* in Fig. 2 has the above form. It finds all information about places in some of the US states, along with their zip codes and weather forecasts. The result set size is scaled by varying the number of selected states. Its dependency graph is shown in Fig. 3.

```
select  gp.TOPLACE, gp.TOSTATE, gz.ZIPCODE, gc.DATE,
        gc.DESCRIPTION
from    GetAllStates gs, GetPlacesInside gp,
        GetInfoByState gi,GetCityForeCastByZip gc,
        getzipcode gz
where   gs.State<'MD' and gi.USState=gs.State and
        gz.zipcode=gp.zip and gc.zip=gz.zipcode and
        gi.GetInfoByStateResult=gz.zipstr
```

Fig. 2. SQL Query1

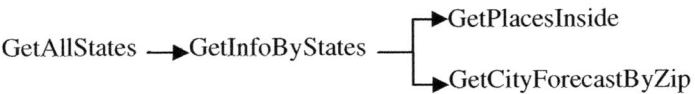

Fig. 3. ADG Query1

For *Query1*, the automatically generated SQL views *GetAllStates*, *GetInfoByState*, *GetPlacesInside*, and *GetCityForeCastByZip* are defined as SQL views that encapsulate four different web service operations from four different service providers with the same names. The SQL view *GetAllStates* presents information of US states as a set of tuples (*state*). The SQL view *GetInfoByState* retrieves all zip codes for a given state as a single comma separated string (*gi.GetInfoByStateResult*). *getzipcode* is a helping function defined in WSMED that extracts the set of zip codes (*gz.zipcode*) given a comma separated string of zip codes (*gz.zipstr*). The SQL view *GetPlacesInside* returns for a given zip code a set of tuples (*ToPlace, ToState, Distance*) where *ToPlace* is a place located within the zip code area, *ToState* is the state where the place is located, and *Distance* is the distance from the place to the origin of the zip code area. The SQL view *GetCityForeCastByZip* reports the weather forecast as a set of tuples (*Date, Description*) for a given zip code where *Date* is date of the forecast, and *Description* is the short description of the forecast. In the above query the SQL view *GetInfoByState* depends on the SQL view *GetAllStates*. The SQL views *GetPlacesInside*, and *GetCityForeCastByZip* depend on the SQL view *GetInfoByState*, while the SQL views *GetPlacesInside*, and *GetCityForeCastByZip* are independent on each other.

2.2   Query2

```
select  gd.Name,gd.LatDegrees,gd.LonDegrees,
        gz.ZIPCODE,gc.DATE,gc.DESCRIPTION
from    GetAllStates gs, GetPlacesInside gp,
        GetInfoByState gi,GetCityForeCastByZip gc,
        getzipcode gz, GetPlaceDetails gd
where   gs.State<'MD' and gi.USState=gs.State and
        gz.zipcode=gp.zip and
        gi.GetInfoByStateResult=gz.zipstr and
        gc.zip=gz.zipcode and gd.Place like '[A-Z]*' and
        gd.Place=gp.TOPLACE and gd.State=gp.TOSTATE
```

Fig. 4. SQL *Query2*

SQL *Query2* in Fig. 4 is a more complex plan than *Query1* having one more dependent SQL view *GetPlaceDetails*, as illustrated by the dependency graph in Fig. 5. It finds all information about places in some of the US states, along with their zip codes, weather forecasts, and geographical positions. In the evaluation the result set size is scaled by varying both the number of selected states and city names.

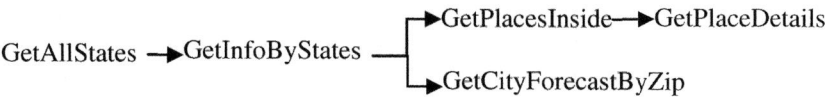

Fig. 5. ADG Query2

For *Query2*, the views *GetAllStates, GetInfoByState, GetPlacesInside,* and *GetCityForeCastByZip* are the same as for *Query1*. The additional SQL view *GetPlaceDetails* returns for a given city and state a set of tuples (*Name,LatDegrees, LonDegrees*) where *Name* is a place located within the city, and *LatDegrees* and *LonDegrees* represents latitude and longitude of the place in degrees, respectively. The SQL view *GetPlaceDetails* depends on SQL view *GetPlacesInside*. *Query2* filters the city name *Place* (*gd.Place like '[A-Z]*'*) since the web service operation *GetPlaceDetails* [3] doesn't support such filters.

3   Query Plans

The WSMED query processor first generates a central preliminary plan containing calls to the web service operations. It is a left-deep tree of executable predicates enumerated from 0 and up. The central plan contains calls to an *apply* operator γ that calls a plan function for a given parameter tuple. The non-parallel query execution plan using γ can be directly interpreted but with very bad performance, since the web service operations are then called in sequence in a single process.

Fig. 6. Central *Plan1C*

In Fig. 6 the central *Plan1C* for *Query1* first calls the plan function that encapsulates the web service operation *GetAllStates* returning a stream of tuples (*state*), which is then selected by the inequality operator '<' at *pos=1*. Each of the *state* tuples are fed to the next plan function, which encapsulates web service operation *GetInfoByState* parameterized by *state* returning a stream of comma separated strings *zipstr*. For each *zipstr* the γ operator calls the user defined helping function *getzipcode* to produce a stream of extracted zip codes (*zipcode*). Then the plan function encapsulating web service operation *GetPlacesInside* is applied on each argument tuple (*zipcode*) to produce a stream of tuples (*toplace, tostate, zipcode*). Finally the plan function for *GetCityForeCastByZip* is applied on each argument tuple (*toplace, tostate, zipcode*) returning as the query result a stream of tuples (*toplace, tostate, zipcode, date, description*).

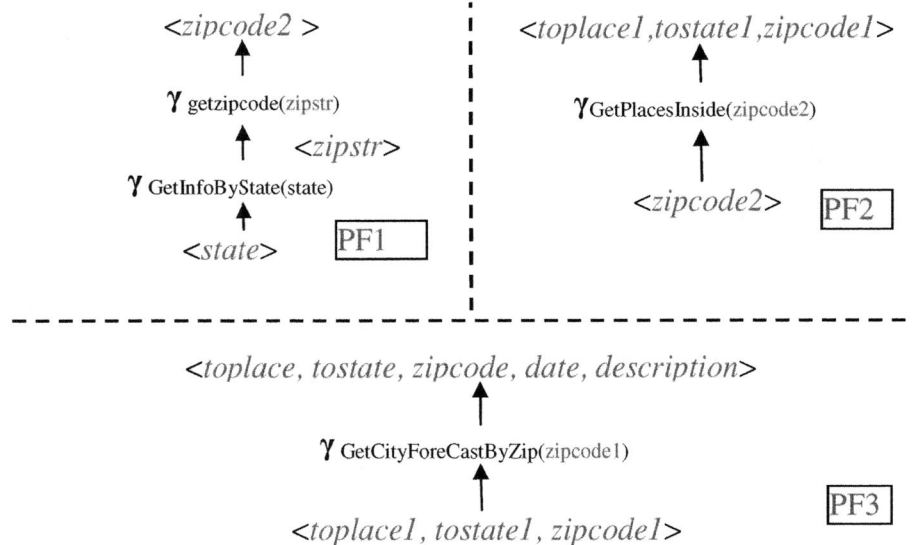

Fig. 7. Dependent plan functions

Plan functions are first generated to encapsulate fragments of the central plan for each web service call. The query processor then transforms the central plan to a plan where the generated plan functions are called in parallel by the *PAP* operator.

For example, Fig. 6 shows the central *Plan1C* for *Query1* and Fig. 7 shows the corresponding three generated query plans functions *PF1, PF2* and *PF3* for *Plan1C*. *PF1* encapsulates the web service operation *GetInfoByState*, and calls the foreign function *getzipcode* to unpack its result. *PF2* encapsulates web service operation *GetPlacesInside* and *PF3* encapsulates *GetCityForeCastByZip*.

In the plan in Fig. 8, the *PAP* operator calls a single plan function at a time. Since the parallelization by the *PAP* operator is based on parameter streams, SQL views encapsulating the web service operations not having input parameters are not considered for parallelization. Therefore the SQL view encapsulating the web service operation *GetAllStates* without input parameters is not considered for parallelization. *PAP* adaptively parallelizes the dependent calls to the plan functions *PF1, PF2* and *PF3* so that they will be executed as a parallel pipeline. This plan is suboptimal since it assumes that all the web service operation calls are considered as dependent on each other. A better plan is taking the independence between the web service operations *GetCityForeCastByZip* and *GetPlacesInside* into account, which will be shown later.

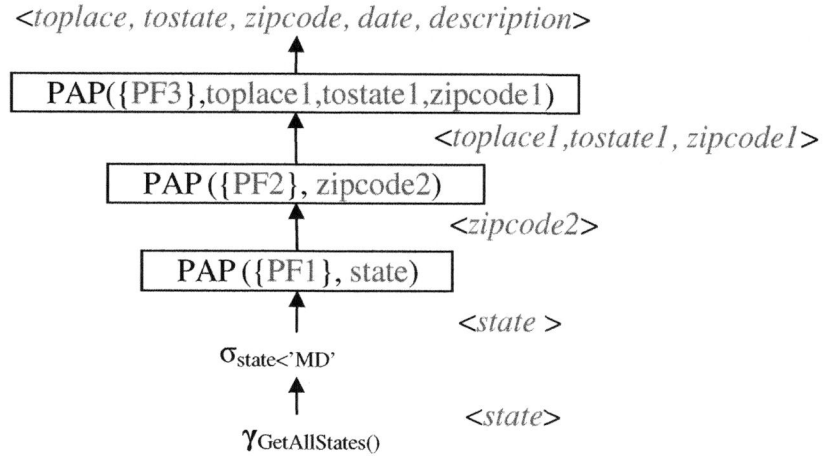

Fig. 8. Dependent adaptive parallel plan *Plan1A*

Fig. 9 shows a process tree for the dependent parallel plan of *Query1* in Fig. 8. Initially a user query is submitted to the coordinator process *q0* for processing. Then the query processor automatically generates the parallel query plan. Once the parallel plan is started, *PAP* will automatically start new parallel processes to dynamically form a process tree at run time. Every query process on each level is connected with several child processes. All processes on the same level execute the same set of plan functions for that level, but with different parameter tuples. On each level always one plan function is applied.

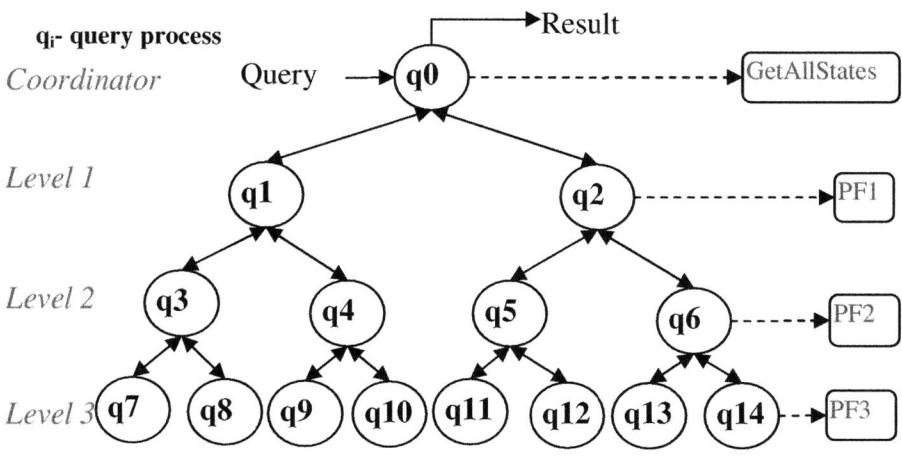

Fig. 9. Adaptive dependent parallel process tree

In Fig. 9, query *q0* is connected with *q1* and *q2*. The execution plan in *q0* calls the non-parameterized web service operation *GetAllStates*, while *PF1* executing in level one calls the web service operation *GetInfoByState* for different states. On level two *PF2* calls the web service operation *GetPlacesInside* for different zipcodes. Finally on level three *PF3* calls the web service operation *GetCityForecastByZip* for different zipcodes.

In this plan the web service operations *GetPlacesInside* and *GetCityForecastByZip* are regarded as dependent on each other. This makes the web service operation *GetCityForecastByZip* be called several times for the same zipcode. Since web service calls have high latency, these redundant calls cause delays. In what follow, we will show how such redundant calls are eliminated for independent web service operations.

In the plan in Fig. 6 the query processor has identified that the web service operations *GetPlacesInside* and *GetCityForeCastByZip* are independent on each other but dependent on *GetInfoByState*. Therefore, in Fig. 11, the plan functions *PF2* and *PF3* in Fig. 7 are replaced by the independent plan functions *PF4* and *PF5* in Fig. 10. They both depend only on *PF1*. For a given zipcode *PF4* calls *GetPlacesInside,* and *PF5* calls *GetCityForeCastByZip*, respectively.

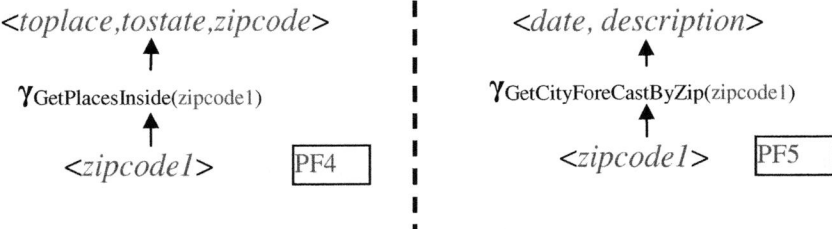

Fig. 10. Independent plan functions

However, in the dependent plan in Fig. 8, for a given Zipcode, *GetPlacesInside* returns more than one place and for each so returned place *GetCityForecastByZip* is called. By contrast in the plan in Fig. 11 no redundant calls are made by considering *PF4* and *PF5* as independent plan functions in a PAP call. Here, the latency of calling *GetPlacesInside* and *GetCityForecastByZip* is reduced by calling them asynchronously and in parallel. Fig. 12 shows the parallel process tree. Contrary to the process tree in Fig. 9, the independent web service operations *GetPlacesInside* and *GetCityForecastByZip* are called in parallel at level two.

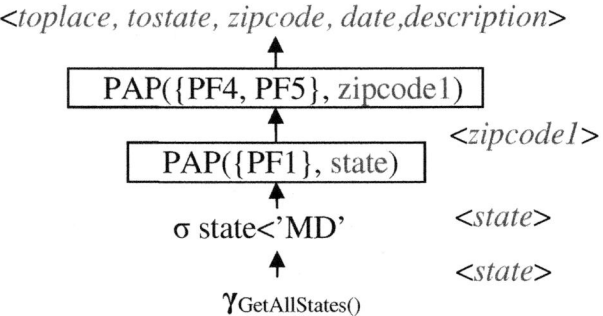

Fig. 11. Dependent and independent adaptive parallel execution plan

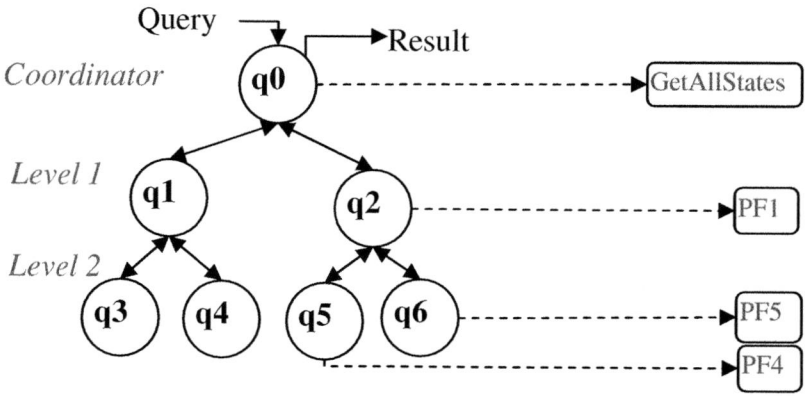

Fig. 12. Adaptive parallel process tree – dependent and independent

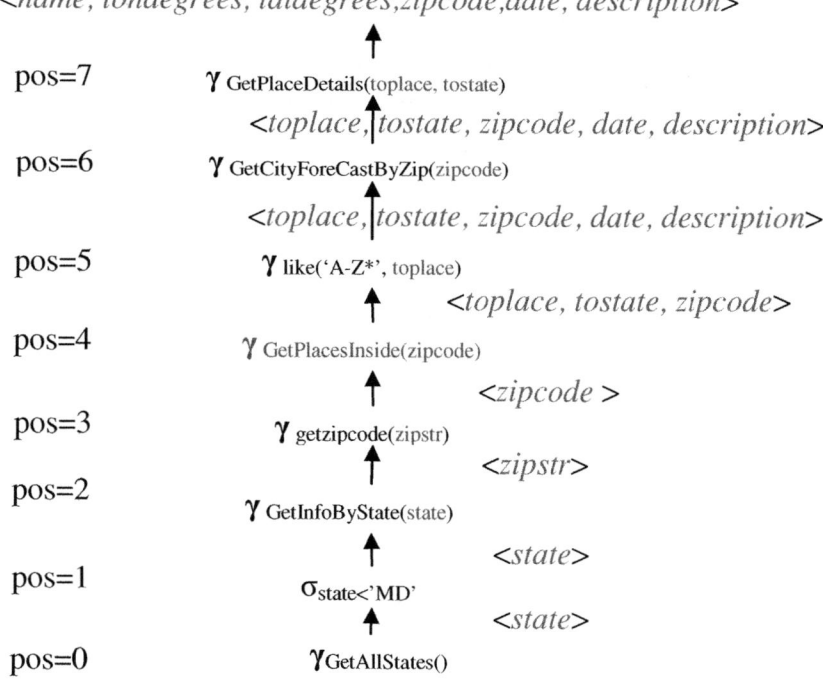

Fig. 13. Central *Plan2*

For the initial central plan, the query processor uses a simple heuristic web service cost model based on the signatures of web service operations assuming that web service operations are expensive. The central plan is then parallelized by generating plans functions and *PAP* calls. The generated central execution plan for *Query2* is *Plan2C* in Fig. 13. One way to parallelize *Plan2C* is to generate the parallel *Plan2aP* in Fig. 15.

Fig. 14. Dependent Plan functions-Query2

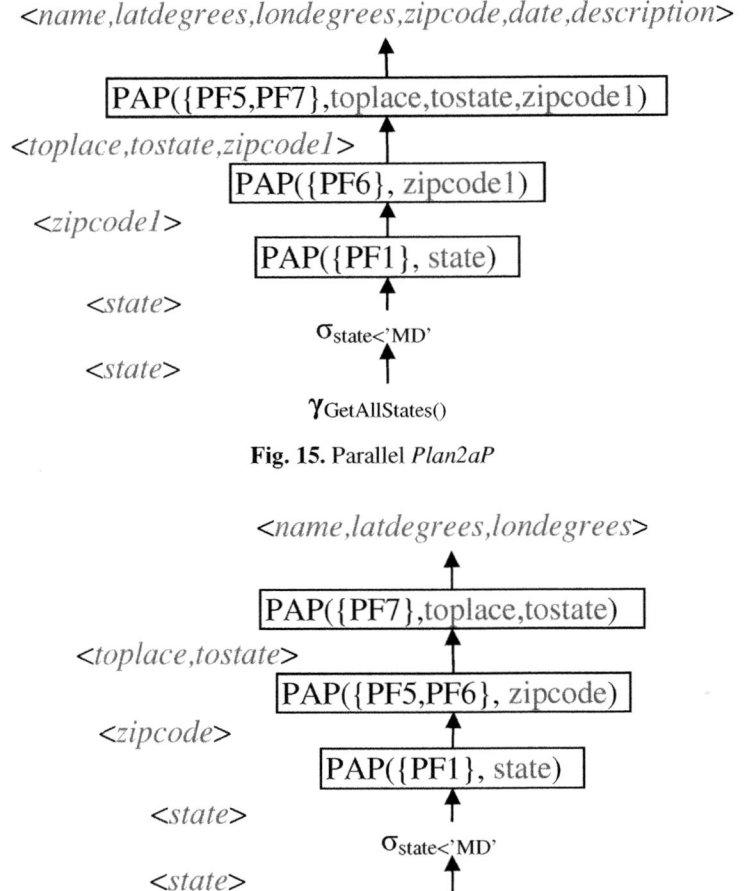

Fig. 15. Parallel *Plan2aP*

Fig. 16. Add stage

Fig. 14 shows the two dependent fragment plan functions *PF6* and *PF7*. However, *Plan2aP* is suboptimal since the plan function *PF5* (Fig. 10) is not dependent on *PF6* but should be applied in parallel with *PF6* instead, as in *Plan2bP* (Fig. 16).

To eliminate such redundant calls, the query processor analyzes the data dependencies in *Plan2C*. In the experimental study section the impact of using these different PAP decompositions is compared. The decomposition basically analyzes data dependencies between plan function calls to push web service operation calls close to where their inputs are produced.

4   The Parameterized Adaptive Parallelization (PAP) Operator

The *PAP* operator calls one or several plan functions in parallel. It has the signature:

```
PAP(Vector of Function vpf, Stream pstream,
    Vector argorder, Vector resorder)→ Stream res
```

PAP takes as arguments a collection of independent plan functions *vpf* along with a stream of parameter values *pstream* to be processed by the plan functions. For each parameter tuple, it starts one process per plan function call in the collection *vpf*. Different plan functions will select different elements from each parameter tuple in *pstream*. The results from the child processes are collected asynchronously and delivered as a stream. The stream of result tuples *res* from PAP are formed by asynchronously combining result tuples from the children. When a child process has delivered all result tuples in a call, then it will send an *end-of-call message* to the parent process where the *PAP* is called and terminate itself, and another child is started asynchronously.

Once started *PAP* dynamically modifies the process tree at run time by locally monitoring the execution time per delivered tuple from each child. If it decreases significantly for some child, new children are added to improve performance.

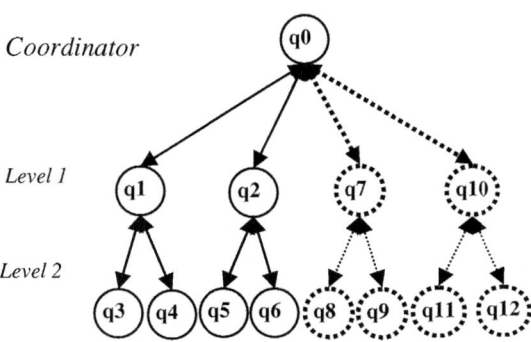

Fig. 17. Add stage

In a process tree, the *fanout* is defined as the number of children processes below a parent query process. A process tree for the execution plan in Fig. 11 is shown in Fig. 12, where every node has fanout two. First, the coordinator *q0* has started two children *q1* and *q2*, each executing the same plan function *PF1*. In the next call to *PAP* the plan functions *PF4* and *PF5* are independent. Therefore, a call to *PAP* is executed in each of *q1* and *q2* with different plan functions *PF4* and *PF5*, respectively. *PAP* in *q1* has created a binary sub-tree with children *q3* and *q4*, while *q2* has the children *q5* and *q6*. The query processes *q3* and *q5* are started with *PF4* while *q4* and *q6* are started with *PF5*.

The pseudo code for PAP is listed in Appendix A. The stages of *PAP* are:

1. It initially forms a process tree by having *fanout* set to the length of *vpf* (Appendix A : lines 1-3,). The fanout is minimally two to ensure parallelism when length of *vpf* is one, as in *q0*. This is called the *init* stage (lines 1-15).
2. A *monitoring cycle* for a non-leaf query process is defined as when *PAP* has received end-of-call messages from all its children and the total number of received result tuples *tc* is at least one (line 49). After the first monitoring cycle when the execution time per tuple for the previous monitoring cycle, *pre_tupt*, is zero (line 54) *PAP* adds *a* new child processes (line 55). Initially *a=2* (line 4) so a binary tree will be first formed in the add stage. This is called the *add* stage. In Fig. 17, the query process *q0* has added two new processes *q7* and *q10* at level 1 compared to Fig. 12.
3. When an added node has several levels of children the init stages of the children's *PAP*s are rerun. That is, *q7* adds *q8* and *q9* while *q10* adds *q11* and *q12* in order to form a new binary sub tree.
4. *PAP* records per monitoring cycle *u* the average computation time t_u (line 51) to produce an incoming tuple from the children. This time is dominated by the latency of the encapsulated web service operations.
   a. If t_u decreases (line 54) more than a threshold (line 9) the add stage is rerun.
   b. If t_u increases (line 56) no more children are added.

The arguments of the plan functions f_i in *vpf* to execute are provided through the input stream *pstream*. For each argument tuple *p* (line 17) in *pstream*, *PAP* starts processes (line 26) executing all f_i in parallel in a round robin fashion. Each *p* provides arguments for all f_i. However, different f_i use different parameter values in *p* and the parameter *argorder* specifies for each f_i how to form the arguments of f_i from *p* (line 25). It is a vector of vectors of argument positions $\{\{a_{ij},...\}...\}$, that specifies per f_i the parameter positions $\{a_{ij},...\}$ to pick from *p*. For example, in Fig. 11 the uppermost call to *PAP* has *argorder* = $\{\{1\},\{1\}\}$ because both *PF4* and *PF5* take as argument the first element of the argument tuple *<zipcode1>*.

Each result tuple *r* (line 42) emitted from *PAP* consists of values r_k. The *PAP* parameter *resorder* specifies how to compute r_k from results of f_i. It is a vector of pairs $\{\{p_{km},c_{km}\}...\}$, that specifies per element position *k* in *r* i) the position p_{km} of the function f_m in *vpf* that computed r_k, and ii) which element c_{km} in the result from f_m to select as r_k. In Fig. 11, *resorder* = $\{\{1,1\},\{1,2\},\{1,3\},\{2,1\},\{2,2\}\}$ specifying the result tuple *<toplace, tostate, zipcode, date, description>*.

A child result tuple is delivered back to the parent (line 31) asynchronously as soon as its plan function f_i has produced a new value. *PAP* stores each received child result in an input buffer (lines 36, 39) per child. When *PAP* has received at least one result tuple (line 41) from every f_i for a given argument tuple p the system will emit one or several result tuples based on the *reorder* and cartesian product (line 42) of the result tuples received from each f_i. Once a child has no more result tuples to emit it terminates (line 43). When the parent receives an *end-of-call* message from a child, it starts another child process (line 26) for the plan function in *vpf* to be called next. It picks its parameter tuple (line 25) from the current argument tuple (line 17).

PAP is terminated when there are no pending parameter tuples (line 62) in *pstream* and no still running children (line 80).

In general *PAP* supports a mixture of parallel paradigms [14]:

— *Pipeline parallelism*: *PAP* calls in parallel several pipelined plan functions encapsulating dependent web service calls.
— *Intraoperation Parallelism*: The plan functions that encapsulate independent web service calls are called in parallel with PAP.

4.1   Experimental Study

Experiments were run under Windows XP on an HP Compaq 530 with a 3 GHz single processor Intel Pentium 4 and 2.5GB RAM. We compared the query execution times for *Query1* using six different strategies:

1. *Dependent (D)*: Strategy D is a naïve dependent strategy as in Fig. 8. This corresponds to the adaptive parallelization with *AFF_APPLYP*[11][12]. All the web service operations in the query are considered as dependent calls, even the independent ones. A new sub-tree is always started with fanout two, which is increased by two by the adaptation.
2. *Dependent with varying minimal fanout (DF)*: Strategy DF measures the impact of varying minimal fanout for a dependent strategy. This is as strategy D, but new sub-trees are started with the same fanout as the current adapted fanout of its siblings. The fanout of the first child of a level is two.
3. *Cached dependent (CD)*: Strategy CD measures impact of caching results from the web service operations for a dependent strategy. It modifies strategy D by caching results of operation calls. For example, in Fig. 8 the result of calling the operation *GetCityForecastByZip* for a given zip code is cached in a main memory table. Whenever the operation *GetCityForecastByZip* is required to be called in the query, the cache table is checked to avoid redundant calls.
4. *CD with DF (CDF)*: The impact of caching combined with varying minimal fanout is investigated for a dependent strategy.
5. *Independent (I)*: Strategy I measures naïve independent calls for the execution plan in Fig. 11. A new sub-tree is always started with fanout equal to the number of plan functions in *vpf* of the *PAP* call. The fanout is two if *vpf* has length one.
6. *Independent with varying initial fanout (IF)*: This is as strategy I, but new sub-trees are started with the same fanout as the current adapted fanout of its siblings.

The experiments were made by scaling *Query1* by selecting an increasing number of states. This produces an increasing number of zipcodes and increases the cardinality of the result.

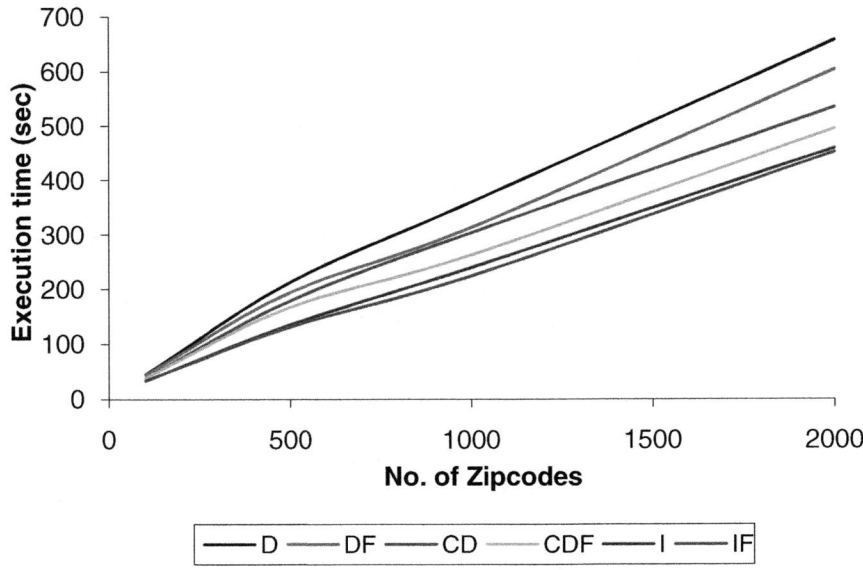

Fig. 18. Experiments with adaptive strategies

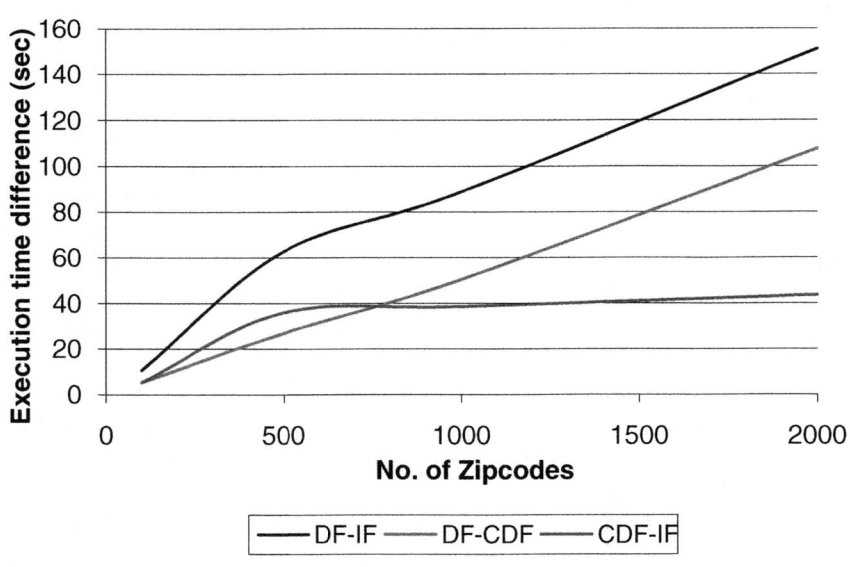

Fig. 19. Relative scalability

Fig. 18 shows that strategy *D* is slowest, and *DF* is somewhat faster. *CD* is even faster, showing that caching is favorable since web service calls incur high latency. *CDF* is even better as it combines caching and adaptive initial fanout. However, even the naïve *PAP* strategy *I* is faster than all variants of the dependent strategies. Strategy *IF* is best. Caching does not pay off for independent strategies, since no redundant calls are made; therefore the combination of caching with *IF* was not measured.

Fig. 19 shows the relative scalability comparing independent and dependent strategies and caching. *DF-IF* plots the performance difference between *DF* and *IF*. It shows that the independent strategy *IF* scales better. Analogously, *DF-CDF* shows for dependent strategies that caching improves scalability. *CDF-IF* shows that the best independent strategy *IF* scales somewhat better that the best dependent strategy *CDF*. However, unlike *CDF*, *IF* requires no extensive memory for caching.

To investigate the impact of adaptation we devised another strategy Non Adaptive Independent(*NAI*). It is similar to *I*, but fanout is fixed to two in all levels of the process tree. Fig. 20 shows that *IF* outperformed *NAI*.

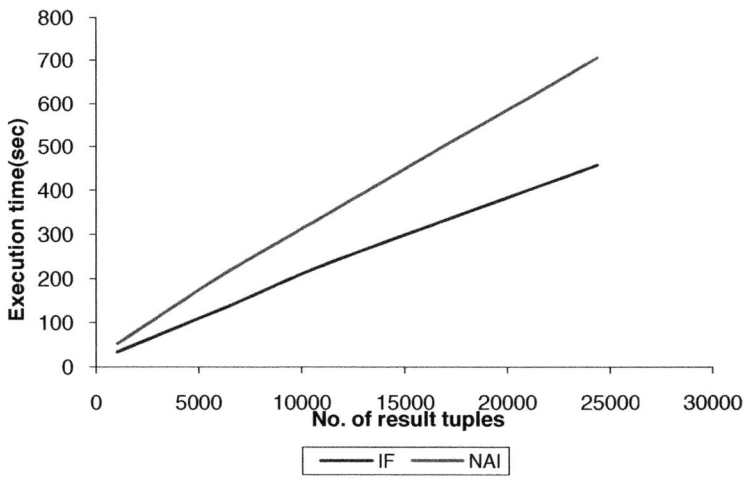

Fig. 20. Impact of adaptation

The performance of different PAP decompositions was investigated by experiments for *Query2* scaled by selecting an increasing number of states and places. Fig. 21 shows that *Plan2bP*(Fig. 16) outperforms parallel *Plan2aP*(Fig. 15). Fig. 22 compares the number of web service calls made by the two different parallel plans.

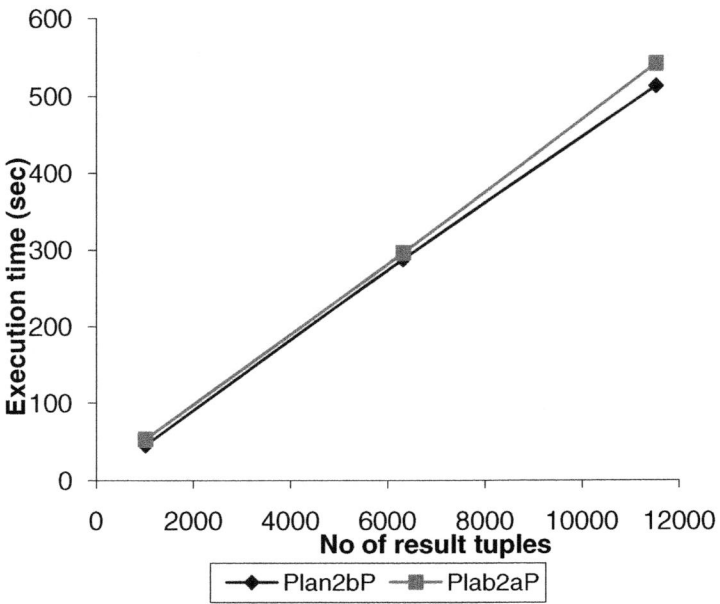

Fig. 21. Impact of central plan execution order

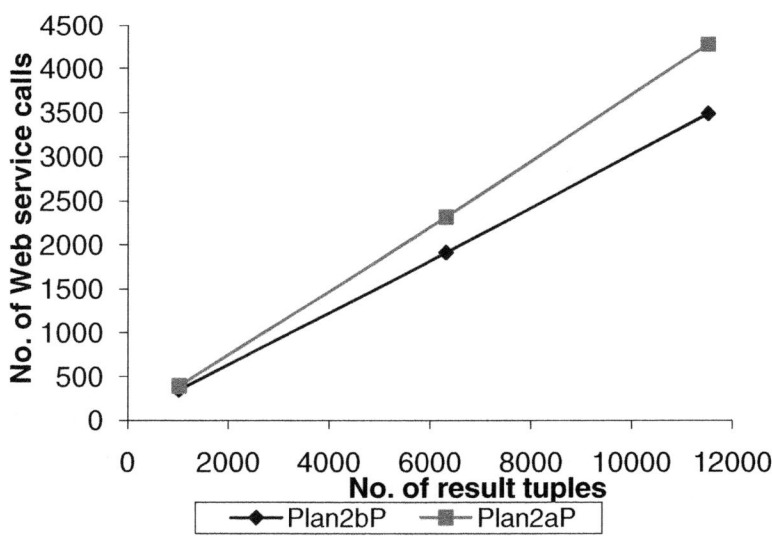

Fig. 22. Number of web service calls-*Pla2aP* Vs *Plan2bP*

5   Related Work

PAP generalizes *AFF_APPLYP* [11][12] by parallelizing both dependent and independent web service operation calls, while *AFF_APPLY* produced a parallelized pipelined plan of only dependent calls.

WSQ/DSQ [9] handles high-latency calls to web search engines by launching asynchronous materialized sub-queries later joined in the execution plan using a special operator without any adaptation. In contrast, WSMED adaptively produces multi-level parallel plans based on streams of parameter tuples passed to parallel sub-plans.

WSMS [13] proposed a cost-based approach for pipelined parallelism among web service operation calls to minimize the query execution time. By contrast, we parallelize adaptively calls to web service operations without any cost model. Furthermore, *PAP* adaptively parallelizes the same web service operation by starting several query processes.

Flow algorithms [8] addressed the problem of minimizing the response time of a multi-way join query calling web service operations using pipelined parallelism. However such algorithms assume the web service operations' cost is known in prior and the produced interleaving query plans are not adaptive. In contrast, *PAP* is optimizing the queries calling web service operations without considering the cost and adaptively parallelizing the calls.

In [10] run time adaptation of buffer sizes in web service calls is investigated, not dealing with adaptive parallelism on web service calls at the client side.

In [1] an approach is described for optimizing web service compositions by traversing ActiveXML documents to select relevant embedded web service calls. It identifies only the independent sub-queries having web service operations and calls them in parallel. The parallelization required a static cost model. By contrast, *PAP* adaptively parallelizes plan functions with both dependent and independent web service calls without any cost model.

Eddies [2] dynamically reorder query processing operators by an n-ary tuple router interposed between *n* data sources and a set of query processing operators. Rather than routing, *PAP* adaptively parallelizes calls to parameterized operators (plan functions) for different parameter values. The purpose of eddies is to avoid dependencies between operators, while the purpose of *PAP* is to speed up calls to individual plan functions. *PAP* complements eddies.

[5] presents algorithms for parallel evaluation of XPath queries. In contrast to such algorithms PAP optimizes the queries without knowing the data distribution and adaptively parallelizing the queries.

Starting query processes with plan functions and the parameter tuple shipping phase of *PAP* has some similarity with the map phase of *MAPREDUCE* [7]. However, *MAPREDUCE* is not dynamically adapting query execution plans as *PAP* and is not streamed.

6   Conclusion

Without any cost knowledge the WSMED query processor automatically and adaptively finds an optimized parallel execution plan calling the queried data providing web services. The algebra operator *PAP* locally adapts the parallel plan by adding children, until no significant performance improvement is measured, based on monitoring the flow between query processes. The operator handles queries where data providing web service operations are called both dependently and independently. *PAP* substantially improves the query performance without any cost knowledge or extensive memory usage compared to other strategies. The measurements are all made with publicly available web service operations.

To lower the number of web service operation calls WSMED decomposes the *PAP* calls to minimize redundant calls. A detailed analysis of the *PAP* decomposition algorithm is future work.

Acknowledgments. This work is supported by the Swedish Foundation for Strategic Research under contract RIT08-0041.

References

1. Abiteboul, S., et al.: Lazy query evaluation for active XML. In: International Conference on Management of Data, pp. 227–238. ACM Press, New York (2004)
2. Avnur, R., Hellerstein, J.M.: Eddies: Continuously adaptive query processing. In: Proceedings of International Conference on Management of Data, pp. 261–272. ACM Press, New York (2000)
3. codeBump-GeoPlaces web service, http://codebump.com/services/PlaceLookup.asmx
4. codeBump-Zipcodes web service, http://codebump.com/services/ZipCodeLookup.asmx
5. Cong, G., Fan, W., Kementsietsidis, A.: Distributed Query Evaluation with Performance Guarantees. In: International Conference on Management of Data, pp. 509–520. ACM Press, New York (2007)
6. CYDNE, http://ws.cdyne.com/WeatherWS/Weather.asmx?WSDL
7. Dean, J., Ghemawat, S.: MAPREDUCE: Simplified Data Processing on Large Clusters. Communications of the ACM, 107–113 (2008)
8. Deshpande, A., Hellerstein, L.: Flow Algorithms for Parallel Query Optimization. In: 24th International Conference on Data Engineering, pp. 754–763. IEEE, Cancun (2008)
9. Goldman, R., Widom, J.: WSQ/DSQ: a practical approach for combined querying of databases and the Web. In: International Conference on Management of Data, pp. 285–296. ACM Press, New York (2000)
10. Gounaris, A., Yfoulis, C., Sakellariou, R., Dikaiakos, M.D.: Robust Runtime Optimization of Data Transfer in Queries Over Web Services. In: International Conference on Data Engineering, pp. 596–605. IEEE (2008)
11. Sabesan, M., Risch, T.: Adaptive Parallelization of Queries over Dependent Web Service Calls. In: Proceedings of First IEEE Workshop on Information & Software as Services, pp. 1725–1732. IEEE computer society (2009)
12. Sabesan, M., Risch, T., Luan, F.: Automated Web Service Query Service. International Journal of Web and Grid Services (IJWGS) 6(4), 400–423 (2010)
13. Srivastava, U., Widom, J., Munagala, K., Motwani, R.: Query Optimization over Web Services. In: Proceedings of Very Large Database Conference, VLDB Endowment, pp. 355–366 (2006)
14. David, T., Clement, H.C.L., Wenny, R., Sushant, G.: High Performance Parallel Database Processing and Grid Databases. Wiley (2008)
15. USZip web service, http://www.webservicex.net/uszip.asmx

Appendix A: PAP Algorithm

	PAP(vpf, pstream, argorder, resorder) $\rightarrow$ result
	input: **vpf** : vector plan functions f_i
	pstream : a stream of parameter values for each plan function f_i in *vpf*
	argorder : specifies how to form the arguments from each argument tuple in *pstream* for each plan function f_i in *vpf*
	resorder : specifies each result r_k in *result* tuple based on a result tuple received from a child.
	output: **result** : Stream of result tuples from children
1	**if** (length of *vpf* >1)
2	*fanout* $\leftarrow$ length of *vpf*
3	**else** *fanout* $\leftarrow$ 2;
4	$a \leftarrow 2$ /* Number of query processes added after each monitoring cycle*/
5	$tc \leftarrow 0$ /* Number of result tuples
6	$ack \leftarrow 0$ /* Number of end-of-call messages*/
7	$t \leftarrow 0$ /*Measured time required to retrieve a tuple (time per tuple)*/
8	*opt*$\leftarrow$false /*Flag indicating whether adaptive expansion of fanout is started (*opt*=true) or stopped (*opt*=false)*/
9	*threshold_value*$\leftarrow$ 0.2 /*Stopping threshold, change in t per cycle*/
10	$exet \leftarrow 0$ /*Execution time of f_i in children processes per cycle*/
11	*bgcount* $\leftarrow -1$ /*Counter of argument tuple p in pstream*/
12	$cid \leftarrow 0$ /*Child identifier for a child query process
13	*bag_buff* $\leftarrow$ empty /*Buffer *bag_buff* keeps *fnvec, carr* for each *bgcount* where *carr* is array of *cids*/
14	*co_buff* $\leftarrow$ empty /*Buffer *co_buff* keeps result tuples and *bgcount* each *cid*/
15	*clist*$\leftarrow$ empty /* List of *cids* of active children*/

16	**while** (*pstream* is not empty)
17	$p\leftarrow$ Get the argument tuple from *pstream*
18	*bgcount*$\leftarrow$ *bgcount+1*
19	$i\leftarrow -1$ /*Position of f_i in *vpf*/
20	*carr* $\leftarrow$ array of *cids* per f_i initially set to 0 /* Each element of the integer array indicates a plan function f_i has executed in a child q_j */
21	*fnvec* $\leftarrow$ array of flags per f_i initiated to *false* /*The boolean array indicates whether a plan function f_i has executed in a child q_j . When q_j produce the first result tuple the respective array element become *true.*/
22	Insert value triples (*bgcount, fnvec, carr*) to the buffer *bag_buff*
23	**while** (i <(length of vpf-1))
24	$i\leftarrow i+1$
25	$ct_i\leftarrow$ Retrieve child argument tuple from *pstream* using *argorder$_i$*

26			$cid_k \leftarrow$ Initialize a query process to execute the plan function f_i with ct_i
27			$carr_i \leftarrow cid_k$ and update the buffer bag_buff with new $carr$.
28			Add cid_k to $clist$
29			$m \leftarrow 0$ /* index to the elements in clist*/
30			**while** ((number of children in $clist$ = $fanout$) **and** ($clist$ is not empty))
31			$res \leftarrow$ a result tuple sent from a child process cid_m
32			$m \leftarrow m+1$ /* increase index to point next element in $clist$*/
33			$tbgcount \leftarrow bgcount$ for child process cid_m from co_buff
34			**if** (res is a valid result)
35			**if** (res is the first result tuple of cid_m)
36			Insert value triples (cid_m, res, $tbgcount$) to the buffer co_buff
37			$fnvec \leftarrow$ get $fnvec$ for $tbgcount$ from bag_buff buffer
38			The array element referred by cid_m in $fnvec$ is set to 1 and update $fnvec$ in bag_buff.
39			**else** Update value triples (cid_m, res, $tbgcount$) to the buffer co_buff
40			$tc \leftarrow tc+1$
41			**if** all $cids$ for $tbgcount$ emit at least one result tuple
42			Construct and emit the result r of PAP from all such $cids$ by retrieving respective res from co_buff. The position of the result values in r determined by $resorder$.
43			**else if** (res is end-of-call message)
44			**if** all $cids$ for $tbgcount$ have sent end-of-call message
45			Remove entries of all $cids$ for $tbgcount$ from co_buff
46			Remove cid_m from $clist$
47			$ack \leftarrow ack+1$
48			$exet \leftarrow exet +$ execution time of cid_m
49			**if** ($ack = fanout$) **and** ($tc > 0$)
50			$pre\_tupt \leftarrow t_u$
51			$t_u \leftarrow (exet / tc)$
52			**if** ($pre\_tupt > 0$)
53			$relative\_error \leftarrow ((pre\_tupt - t_u)/(pre\_tupt))$
54			**if** ((($threshold\_value < relative\_error$) **and** (**not** opt)) **or** ($pre\_tupt = 0$))
55			$fanout \leftarrow fanout + a$
56			**else if** ($threshold\_value >= relative\_error$)
57			$opt \leftarrow$ true
58			**if** ($ack = fanout$)
59			$ack \leftarrow 0$; $exet \leftarrow 0$; $tc \leftarrow 0$;
60		**end while**	
61	**end while**		
62	**end while**		

63	m←0 /* index to the elements in *clist*/			
64	**while** (*clist is not empty*) /* some child process left to be finished * /			
65		*res* ← the result tuple sent from a child process cid_m		
66		m←m+1 /* increase index to point next element in *clist*/		
67		*tbgcount*← the *bgcount* for child process cid_m from *co_buff*		
68		**if** (*res* is a valid result)		
69			**if** (*res* is the first result tuple of cid_m)	
70				Insert value triples (cid_m, *res*, *tbgcount*) to the buffer *co_buff*
71				*fnvec*← get *fnvec* for *tbgcount* from *bag_buff* buffer
72				The array element referred by cid_min *fnvec* is set to 1 and update *fnvec* in *bag_buff*
73			**else** Update value triples (cid_m, *res*, *tbgcount*) to the buffer *co_buff*	
74			**if** all *cids* for *tbgcount* emit at least one result tuple	
75				Construct and emit the result of PAP from all such *cids* by retrieving respective *res* from *co_buff*. The position of the result values in *r* determined by *resorder*.
76		**else if** (*res* is end-of-call message)		
77			Remove cid_m from *clist*	
78			**if** all *cids* for *tbgcount* have sent end-of-call message	
79				Remove their respective entries in *co_buff* buffer
80	**end while**			

A Pattern-Based Approach
for Efficient Query Processing over RDF Data

Yuan Tian[1], Haofen Wang[1], Wei Jin[2], Yuan Ni[3], and Yong Yu[1]

[1] Shanghai Jiao Tong University, 800 Dongchuan Rd., Shanghai, China
{tian,whfcarter,yyu}@apex.sjtu.edu.cn
[2] North Dakota State University, 1340 Administration Ave, Fargo, ND, US
wei.jin@ndsu.edu
[3] IBM China Research Lab, Beijing, China
niyuan@cn.ibm.com

Abstract. The recent prevalence of Linked Data attracts research interest towards the efficiency of query execution over the web of data. Search and query engines crawl and index triples into a centralized repository and queries are executed locally. It has been shown in various literatures that the performance bottleneck of large scale query execution lies in joins and unions. Based on the observation that a large part of join operations result in a much smaller binding set which can be precomputed and stored, we propose to augment RDF indexes to store the bindings of complex patterns and exploit these patterns to enhance performance. In addition to the index, we also introduce two strategies of selecting these patterns: one depends on developed heuristic rules and the other employs query history to optimize time-space ratio. Our empirical study demonstrates the proposed pattern index outperforms traditional triple index by up to three orders of magnitude while keeping the overhead low.

1 Introduction

Recently, the Linking Open Data movement provides a light-weight approach of publishing semantic data which in turn triggers the explosion of interconnected data over the Web. The problem of retrieving information from this large body of graph data has become pressing for the semantic web community. Though various distributed or federated query engines have been proposed, the mainstream of Linked Data query engines follows the paradigm of traditional search engines, where triples are crawled from the web and hence indexed and stored in a centralized repository. In this paper, we mainly focus on the performance issue of this type of query engines.

Prevailing RDF storage techniques such as Jena [17], Sesame [4], and RDF-3X [9,10] index statements by triple patterns. RDF query engines decompose SPARQL queries into Basic Graph Pattern (BGP) matching problems. And the bindings to each BGP can be retrieved by self joins. It may incur performance degradation when the predicate binds to a large amount of triples, i.e. a less selective predicate. For instance, there are more than 180,000 triples with predicate

A. Hameurlain et al. (Eds.): TLDKS V, LNCS 7100, pp. 70–90, 2012.

"dbpedia:genre" in the DBPedia dataset. Joining a pattern with triple pattern (?x, dbpedia:genre, ?y) is formidably time consuming since fetching such a large amount of triples from disk is extremely expensive in terms of disk IO, not to mention the join operation. Property tables [18,7] were proposed to speed up query performance. They are however not adaptable to relations among resources and require users to manually specify the property table.

In this paper we propose a pattern-based approach of storing RDF data, extending index entries from triple patterns to more complex BGPs. In our approach, the bindings of selected BGPs are precomputed and indexed which can be leveraged to reduce the number of join operations. We observed that queries containing more relations and properties are more selective, which implies fewer bindings. Materializing their bindings provides intermediate results that can be directly fetched with much lower cost than computing them again. In addition, we use a dynamic programming optimizing algorithm that leverages the materialized bindings to reduce the execution cost. To achieve the best overall performance gain, We compare two pattern strategies for generating BGP materialization. One strategy applies cost model heuristics and the other mines frequent patterns.

We list the challenges we are facing as follows:

1. In order to perform fast lookups of stored bindings, efficient conversion between BGPs and index keys is desired.
2. To improve performance, query planning algorithms should be modified accordingly to leverage indexed BGPs.
3. BGPs whose bindings are precomputed and indexed should be carefully selected to achieve best time-space efficiency.

Our main contributions are

1. We extend triple pattern index to BGP index while keeping lookup operations efficient. Triple indexes can be viewed as special cases in our approach.
2. We devise a query optimization algorithm which takes advantage of the indexed BGPs that significantly reduces the number of join operations.
3. We show two different strategies of selecting BGPs, respectively based on heuristic rules and pattern frequencies, and analyze their advantages and disadvantages.
4. We prove by extensive empirical studies that our approach outperforms triple indexes by up to 3 orders of magnitude. And we investigate the relationship between time and space to show the overhead of our approach.

The outline of this paper is as follows. In Section 2, we introduce preliminary concepts and define our problem. And Section 3 proposes hashed pattern and the index structure. Following that, we introduce in Section 4 an approach to incorporate pattern index with existing query planning algorithms. Section 5 describes the two strategies of selecting BGPs in indexing. We show in Section 6 the performance increase of our approach and analyze its scalability and time-space relation. Finally, we list related work in Section 7 and conclude in Section 8.

2   Problem Definition

We begin our problem statement by introducing the preliminary concepts, including the data graph, the pattern, the query and the binding.

2.1   Data Graph

Resource Description Framework(RDF) is officially adopted by W3C [3] as a way of conceptual modeling. The underlying idea is to make statements about resources in the form of subject-predicate-object triples. The subject and object refer to two resources and the predicate between them advocates their relationship. For example, "SJTU is located in Shanghai" could be written into an RDF statement $(SJTU, isLocatedIn, Shanghai)$. A collection of RDF triples constitutes a labeled, directed multi-graph. The concept of data graph here corresponds to this graph.

Fig. 1 exemplifies an RDF graph. We follow the notations and definitions in [3] throughout this paper. The ellipses represent resources while rectangles represent literals.

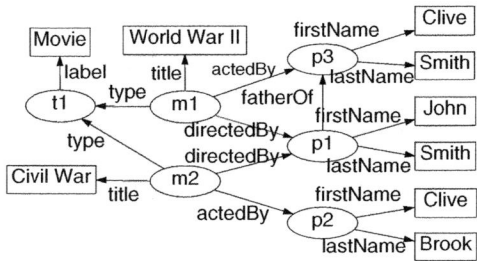

Fig. 1. An example of RDF graph

We can further simplify this model by modeling the properties of a resource as its labels. Stated in detail, each literal and its connecting edge are eliminated from the original multi-graph and the concatenation of each property's name and value is added upon its corresponding resource. By treating literals as graph labels, we can easily index and perform quick look up on these properties. This model has the same expressive power as the original one. We call this graph the data graph. In the rest of this paper, we always denote the data graph as φ.

Definition 1 (Data Graph). *A data graph is a directed and labeled multi-graph $\varphi = (V_\varphi, E_\varphi, l_\varphi)$ where V_φ denotes the set of vertices and $E_\varphi \subseteq V_\varphi \times V_\varphi$ the set of edge. l_φ is the label function which maps nodes and edges in φ to sets of strings.*

Fig. 2 is the data graph of Fig. 1. We shall see the relationship between an RDF graph and its data graph. Take $t1$ for instance. It has a property "label" with value "Movie", which is modeled as the label "label:Movie" depicted in Fig. 2, where ":" is a delimiter. Fig. 1 and Fig. 2 have identical semantic meanings.

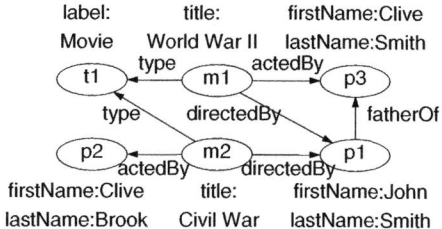

Fig. 2. An example of data model

2.2   SPARQL and Basic Graph Pattern

SPARQL queries are essentially graph pattern matching problems. More complex query patterns can be decomposed into a combination of smaller patterns [11] among research towards which the Basic Graph Pattern (BGP) is highly emphasized. In the representation throughout this paper, a BGP is defined as follows.

Definition 2 (Basic Graph Pattern). *A query pattern is a connected labeled directed multi-graph $p = (V_p, E_p, c_p)$, where V_p specifies a set of vertices and E_p a set of edges. c_p is a mapping from $V_p \cup E_p$ to strings and $\varnothing$.*

Inside the box below is an example of SPARQL query. When translated into graphs, it can be represented as that in Figure 3. Every variable that appears in the SPARQL query corresponds to a node in Fig. 3. Each statement linking two resources by a predicate is represented by a labeled edge. In our example, "?w prop:actedBy ?x" is an edge labeled "prop:actedBy" linking nodes ?w and ?x. On the other hand, statements that relate a resource and a literal are represented by labels on nodes. For instance, the concatenation of the literal "Movie" and the predicate "rdf:label" is added upon node ?z.

```
PREFIX prop:<http://dbpedia.org/property/>
PREFIX res:<http://dbpedia.org/resource/>
SELECT ?x ?y
WHERE {
        ?w prop:actedBy ?x .
        ?w rdf:type ?z .
        ?w prop:directedBy ?y .
        ?y prop:fatherOf ?x .
        ?z rdf:label "Movie" .
        ?w rdf:label "War" .
        ?x prop:firstName "Clive" .
        ?y prop:firstName "John" .
}
```

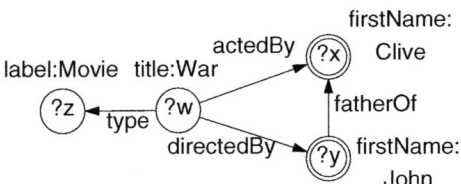

Fig. 3. An example of a query graph

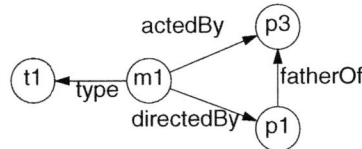

Fig. 4. A solution graph of Fig. 3

2.3   Solutions and Bindings

In graph representation, answers to queries are represented as subgraphs that fulfills the given conditions. If a subgraph of data graph satisfies the condition embodied in a pattern, we call the subgraph binds to the pattern. We formally define solutions and bindings as follows.

Definition 3 (Solutions and Bindings). *The solution set between a pattern* $p = (V_p, E_p, c_p, h_p)$ *and the data graph* $\varphi = (V_\varphi, E_\varphi, l_\varphi)$ *is the set* Ω *of all injective functions* $f : V_p \rightarrow V_\varphi$, *one of which is called a solution of p, such that*

1. *For all* $u \in V_p$, *it holds that* $c_p(u) \in l_\varphi(f(u))$ *or* $c_p(u) = \varnothing$
2. *Every* $e = (u, v) \in E_p$ *agrees with* $e' = (f(u), f(v)) \in E_\varphi$ *and* $c_p(e) \in l_\varphi(e')$

In a match f, if $f(u) = v$ *where* $u \in V_p$ *and* $v \in V_\varphi$, *we call v a binding of u.*

As well as giving a query graph, users are asked to select a set of nodes whose bindings are returned. The bindings of each selected variable for each solution in Ω are returned. From Ω users can get the subgraphs of the data graph φ that match the given pattern.

Fig. 3 shows a sample query graph. Nodes $?x$, $?y$ are selected to be returned. This is meant to retrieve a pair of father and son, respectively named "John" and "Clive", where the son acted in a movie with name "War" directed by the father. With $?z$ instantiated by $t1$, $?w$ by $m1$, $?x$ by $p3$, and $?y$ by $p1$, as is shown in Fig. 4, the relations between these nodes and the labels satisfy the template. Fig. 4 is a subgraph of Fig. 4 which matches the template in Fig. 3. Double-circle nodes are the ones selected such that the pairs of instantiations of x and y should be returned as a result.

3   Pattern Index

In this section, we present our pattern index in three aspects: data structure, hashed patterns and pattern encoding. In a pattern index, the bindings of selected BGPs are precomputed and stored. These materialized results can be employed to enhance the query performance. Extending index to support BGPs poses two challenges : (1) The number of BGPs is large; (2) The index design of pattern index is subtle. We will address these two issues in the following subsections.

3.1   Hashed Patterns

We propose hashed patterns to reduce the number of index entries. The number of large BGPs, i.e. with more nodes and edges, will be much greater than smaller ones. Indexing them all will result in a clumsy index file that results in poor lookup performance. However, the number of bindings of a BGP decreases dramatically as its complexity increases. Hence, we combine several large BGPs into the same hash bucket to strike a balance between the binding number of each bucket and the total index size. These two phenomenon lead us to combining several complex patterns into the same hash bucket such that we might be able to strike a balance between the number of bindings per pattern and the total number of possible patterns.

Definition 4 (Hashed Pattern). *A hashed pattern is a connected directed graph $p = (V_p, E_p, c_p, h_p)$, where V_p specifies a set of vertices and E_p a set of edges. c_p is a mapping from E_p to the union of all strings and $\varnothing$. $h_p : V_p \mapsto \mathbb{Z} \cup \{\varnothing\}$ is the hashing constraint on a node.*

The solutions to a hashed pattern h are the union of the solutions of the BGPs whose hashed pattern is h.

Definition 5 (Solutions of Hashed Pattern). *The solution set between a hashed pattern $p = (V_p, E_p, c_p, h_p)$ and the data graph $\varphi = (V_\varphi, E_\varphi, l_\varphi)$ is the set Ω of all injective functions $f : V_p \to V_\varphi$, one of which is called a solution of p, such that*

1. *Every $e = (u, v) \in E_p$ agrees with $e' = (f(u), f(v)) \in E_\varphi$ and $c_p(e) \in l_\varphi(e')$*
2. *For $\forall u \in V_p$ there exists $s \in l(f(u))$ such that $h_p(u) = hash(s)$ if $h_p(u) \neq \varnothing$*

where hash is a hash function that maps strings to integers.

Fig.5 is an example of hashed patterns. The two graphs in the dashed boxes on the left are two patterns and the one on the right is a hashed pattern. The two graphs in the middle are subgraphs of the data graph in Fig.2. The dotted lines represent binding relationships. Let's suppose our hash function maps "title:War" to 4, "lastName:Smith" to 10, and "lastName:Brook" to 10. According to our definition of hashed solutions, the node with hash label 4 is mapped to $m1$ and $m2$ while the one with label 10 to $p3$ and $p2$. Both subgraphs in the middle are solution bindings in RDF graph.

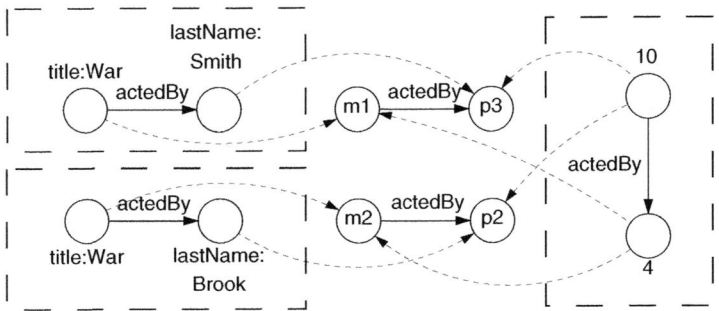

Fig. 5. An example of hashed patterns

3.2 Index and Storage Structure

Usually, triples are stored as records in a heap file. We follow this design: each record stores a solution of some BGP g or some hashed pattern h. The i-th element in the record holds the ID of the resource or the variable that binds to the i-th variable. The solutions can be indexed by various data structures such as B-tree or Inverted Index as long as their keys can be of string type. The distinction between various indexes is not our focal point in our paper so we mainly focus on Inverted Index. We propose to use a mapping function that establishes a one-to-one mapping between BGPs and hashed patterns to strings. When looking up the bindings of a pattern, it is first converted into string and the entry with this string as key is fetched.

Fig. 6(b) shows an example of the heap file and the index for the BGPs in Fig. 6(a). Ordered index is adopted in this example. The left box in Fig. 6(b) is

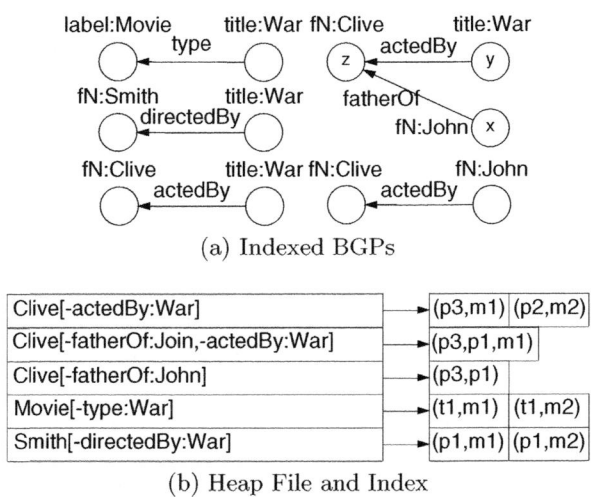

Fig. 6. An example of the hashed patterns

the index with each line representing a key. Each key points to a starting address of its bindings in the heap file. For example, pattern "*[+type::*[+label::Movie]]" binds to pairs $(t1, m1)$ and $(t2, m2)$.

3.3   Encoding of Patterns

Two issues have to be addressed before we can index patterns. The first is in order to transform graphs into existing index structures, such as inverted list, hash buckets or B+tree, we have to convert patterns into ordered index entries. The second is to enable quick judgment of pattern equivalence. Because isomorphism is transitive, two isomorphic patterns always have the same matching set and they are hence regarded as equivalent. These two problems can be simultaneously solved by mapping patterns onto strings in our approach. We call this mapping *encoding* and it should satisfy at least two requirements: (1) Two isomorphic graphs are mapped to the same index entry for the reason that they always have the same matching set. (2) The conversion should be efficient so as to achieve high performance.

As the determination of graph isomorphism is NP Complete [8], we restrict our patterns to be tree-shaped, i.e. acyclic if regardless of the edges' directions. Before we go to our description of encoding, we give the iterative definition of sets D_i of a pattern $p = (V_p, E_p, c_p, h_p)$ as

1. D_1 is the set of all nodes in V_p whose sum of the in-degree and out-degree is 1. This set is non-empty because p is of tree-shape.
2. D_{i+1} ($i \geq 1$) is the set of all nodes $u \notin \bigcup_{k=1}^{i} D_k$ subject to for all $(u, v) \in E_p$ and $(v, u) \in E_p$, $v \in \bigcup_{k=1}^{i} D_k$.

Depending on sets D_i, we can define the encoding function. The encoding function $code(u)$ of $u \in D_1$ is $code(u) = n(u)$ and that of $u \in D_i, i > 1$ is

$$code(u) = n(u)[dir(e_{\{u,w_1\}}) :: code(w_1), ..., dir(e_{\{u,w_n\}}) :: code(w_n)] \quad (1)$$

where each w_i is in $\bigcup_{k=1}^{i-1} D_k$ and is adjacent to u, which are called u's children. It holds that $code(w_i) \preceq_{lexico} code(w_{i+1})$. In the above formula, $e_{\{u,w_i\}}$ represents the edge between u and w_i and $dir(e_{\{u,w_i\}}) = \begin{cases} \text{"+"} & \text{if } e_{\{u,w_i\}} = (u, w_i) \\ \text{"−"} & \text{if } e_{\{u,w_i\}} = (w_i, u) \end{cases}$ and

$n(u) = \begin{cases} c_p(u) & \text{if } c_p(u) \neq \varnothing \\ h_p(u) & \text{if } c_p(u) = \varnothing \text{ and } h_p(u) \neq \varnothing \\ \text{"*"} & \text{if } c_p(u) = \varnothing \text{ and } h_p(u) = \varnothing \end{cases}$. The order $\prec_{lexico}$ is the

lexicographic order of strings.

Let m be the maximum value of i such that $D_i \neq \emptyset$. If $D_m = \{u\}$, we define r as u. If $D_m = \{u, v\}$, we define r as u such that $(u, v) \in E_p$. We can therefore define the encoding of pattern p as $code(r)$. It can be proved that this encoding satisfies our requirements and the calculation of encoding can be done within $O(n)$ time (Due to space limitation, the proof is omitted here.). We consequently assign an order $\prec_{V_p}$ to the nodes such that for all nodes in equation (1) it holds that (1) $u \prec_{V_p} w_i$; (2) $w_i \prec_{V_p} w_j$ if $i < j$; (3) supposing $\{w_i\}$ are u's children and $\{w_j'\}$ are v's children and $u \prec_{V_p} v$, it satisfies $w_i \prec_{V_p} w_j'$ for all i and j. Let us illustrate the encoding by an example.

Take the pattern in the top right corner of Fig.6(a) for instance. According to the definition, we can conclude that $D_1 = \{x, y\}$ and $D_2 = \{z\}$. For nodes in D_1, we have $code(x) =$ "$John$" and $code(y) =$ "War". And for the node z in D_2, it is connected to nodes y and x in D_1 and it holds that $code(x) \prec_{lexico} code(y)$, so that $code(z) =$ "$fN : Clive[-fatherOf :: fN : John, -actedBy :: title : War]$". Finally, because of $D_m = D_2 = z$, the encoding of this pattern would be "$fN : Clive[-fatherOf :: fN : John, -actedBy :: title : War]$". The encoding of hashed patterns is similar except for substituting the labels with hash numbers in the appropriate places.

4   Query Evaluation

In this subsection, we will give a brief review on the query evaluation phase of traditional RDF store. Furthermore, supposing that our indexed patterns are already selected, we will discuss how we evaluate a query in an effective way. Before starting our statement, we introduce concepts of subpattern and covering.

Subpattern relationship is a partial order assigned to patterns. A query pattern p_1 is a subpattern of p_2, denoted as $p_1 \prec_p p_2$, if there exists a mapping $m : V_{p_1} \rightarrow V_{p_2}$, such that (1) For all $u \in V_{p_1}$ with $c_{p_1}(u) \neq \varnothing$, it holds that $c_{p_1}(u) = c_{p_2}(m(u))$; (2) For all $u \in V_{p_1}$ with $c_{p_1}(u) = \varnothing$ and $h_{p_1}(u) \neq \varnothing$, it satisfies $h_{p_1}(u) = hash(c_{p_2}(m(u)))$; (3) For all $(u, v) \in E_{p_1}$, it agrees with $(m(u), m(v)) \in E_{p_2}$ and $c_{p_1}(e) = c_{p_2}(e')$ where $e = (u, v)$ and $e' = (m(u), m(v))$.

A query pattern can be divided into two subpatterns that *cover* it. We say patterns p_1 and p_2 cover p, if there exists a injective function $m : V_{p_1} \cup V_{p_2} \rightarrow V_p$, such that (1) Each $u \in V_{p_i}$ satisfies $c_{p_i}(u) = c_p(m(v))$ or $c_{p_i}(u) = \varnothing$, where $i \in \{1, 2\}$; (2) For each $u \in V_p$, there exists $v \in V_{p_i}$ such that $m(v) = u$ and $c_{p_i}(v) = c_p(u)$, where $i \in \{1, 2\}$; (3) For each $e = (u, v) \in E_p$, there exists $x, y \in V_{p_i}$ with $m(x) = u, m(y) = v$ and $e' = (x, y) \in E_{p_i}$ and $c_{p_i}(e) = c_p(e')$, where $i \in \{1, 2\}$. The function m is called a *covering*.

4.1   Pattern Tree

RDF query engines, no matter whether it is based upon SQL databases or is itself a native RDF store, convert queries written in descriptive query languages, i.e. SPARQL, into executive query plans. Since descriptive languages carry no information on how the answer is computed, a planner is designed to determine the most efficient computational steps taken, a sequence of which is called a plan. Query plans can be represented as trees. Every tree node is an operator, which absorbs input from its children and yields output to its parent. We formalize different pattern matching schemes as pattern trees.

Assume that the solution sets of two subpatterns are already known, the solution set of their super pattern being covered can be calculated by a join of them on their common nodes, i.e. node pairs $(u, v), u \in p_1, v \in p_2$ with

$m(u) = m(v)$ where m is the covering. The proof is straight forward. As for covering m, because of $m(u) = m(v)$, we have $f_p(m(u)) = f_p(m(v))$. Every solution f_p of p is calculated from two solutions f_{p_1} and f_{p_2} of its subpatterns, so that $f_p(m(x)) = f_{p_i}(x)$ for all nodes $x \in p_i$. Thus for common nodes u and v, it requires that $f_{p_1}(u) = f_{p_2}(v)$. The strict semantic meaning of join can be found in [2,12]. In this way, by iteratively dividing the query graph into patterns of smaller granularity and by joining the solution sets of leaf nodes together along the path to the root, we can finally get the result. Our pattern selection strategies to be detailed in Section 3 guarantee that directly retrieving the solution set of an indexed pattern consumes less time than other possible execution plans.

However, we can easily observe that the division is not unique. Different divisions can be represented by different pattern trees. A *pattern tree* is a tree T where each node represents a pattern and the solution set of the pattern represented by a node is calculated by a join of the solution sets of its children.

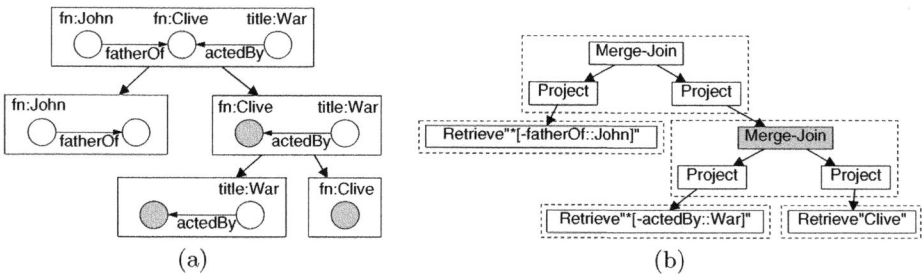

Fig. 7. An example of the pattern tree and the execution plan

Shown in Fig.7(a) is a pattern tree. Each box is a pattern tree node and the graph inside the box denotes the pattern represented by the node. As we shall see, every pattern of a non-leaf node is covered by the patterns represented by its children.

A pattern tree can be directly converted into an execution plan. An execution plan is usually represented by a tree, indicating the input-output relation among the operators [14,6]. A non-leaf query pattern tree can be interpreted as a projection that eliminates the information about the solution of query nodes that are neither to be joined on nor to be returned, followed by an equal-join on the common query graph nodes taking the projected results as input. Various join types, such as merge-join, hash-join and index-join, could be chosen according to the feature of its input[14].

Fig. 7(b) is the transformed execution plan of Fig. 7(a). For simplicity, the prefix of node label is omitted and the example uses unhashed patterns. Solid boxes denote operators which are the nodes in the plan, inside which their types are shown. The solid boxes inside the same dotted box are transformed from the same pattern tree node. Take the shaded equal-join node in Fig. 7(b) for instance.

The patterns of its left child "*[actedBy::War]" and its right one "Clive" are overlapped on the shaded nodes in Fig. 7(a). Hence, the binding IDs of this node get equally joined.

4.2   Query Planning

Query planning is the procedure of choosing the optimized execution plan in the search space [6]. In order to perform quantitative assessment on the amount of effort taken to execute the query according to a plan, we need to establish a cost model. Here we incorporate the optimization of the pattern tree with that of the plan. A *cost model* is actually a cost function $T \rightarrow \mathbb{Z}$, where T is all execution plan in the search space. The cost model is determined by the design of its operators, histograms collected, etc. More details on cost estimation can be found in the publications mentioned in [6].

Despite different kinds of cost estimation, it always holds that if r is not indexed,

$$cost(T_r) = \sum_{\text{all children } c_i \text{ of } r} cost(T_{c_i}) + w(\{c_i\}) \tag{2}$$

where T_r is a plan answering the query pattern r, $\{c_i\}$ is the set of covering subpatterns of r and T_{c_i} is a subtree of T_r answering the pattern c_i. In the equation, $w(\{c_i\})$ is the effort paid to calculate the solution set from the matches of c_is. The main idea of the above equation is that $w(c_i)$ only depends on the set $\{c_i\}$ and has no relation with T_{c_i}. Intuitively, no matter what T_{c_i} is, the matches of c_is will not be altered and hence the effort paid(w) to generate r's answer from these results will be the same. The selection of $\{c_i\}$ for r is actually the selection of r's children nodes in the pattern tree.

Given equation (2), dynamic programming can be adopted to find the plan with the minimum cost. Actually, this optimization problem is to find the pattern tree that will lead to the best plan.

Theorem 1. *The execution plan with minimum cost to match query pattern p, denoted as $mincost(p)$, can be given as follows.*
If p is indexed,

$$mincost(p) = \begin{array}{c} \text{Equal-Join} \\ \diagup \qquad \diagdown \\ \text{Project} \qquad \text{Project} \\ | \qquad\qquad | \\ mincost(p_1) \qquad mincost(p_2) \end{array}$$

where

$$p_1, p_2 = \min_{p_1,p_2} \{cost[mincost(p_1)] + cost[mincost(p_2)] + w(\{p_1,p_2\})\}$$

and it holds that p_1 and p_2 covers p.
Otherwise, $mincost(p) = $ Retrive p.

Proof. From (2), it holds that

$$\min_{T_p}\{cost(T_p)\} = \min_{p_1,p_2}\{\min_{T_{p_1},T_{p_2}}[cost(T_{p_1}) + cost(T_{p_2})] + w(\{p_1,p_2\})\}$$
$$= \min_{p_1,p_2}\{[cost[mincost(T_{p_1})] + cost[mincost(T_{p_2})] + w(\{p_1,p_2\})\}$$

For indexed ones, a single retrieval operator is the best out of all execution plans as we have stated before.   □

The search space of our algorithm is limited to PJ trees, where a pattern's result is generated by the projections of its covering subpatterns followed by a join. This resembles the SPJ plan [6] firstly proposed in System-R except that in our plan the selects are pushed downwards to leaves and incorporated into the retrieval operator. According to [12], the join operators satisfy the association rule. The search space can be further limited to left-deep plans [14] without affecting the performance. Finally, the best execution plan answering query q is $mincost(p_q)$, where $p_q = (V_q, E_q, c_q, h_\varnothing)$ and $h_\varnothing(u) = \varnothing$ for all u.

5   Pattern Selection

As stated before, pattern-based RDF index challenges us with two problems, the enormous amount of possible complex patterns and the sparsity of bindings. Though obviously materializing bindings obtains performance gain, patterns even if they are hashed are still large in quantity and some of them are seldom accessed as intermediate results. Thus it is neither feasible nor wise to store them without selection. To strike a balance between space and time, different metrics could be taken to assess or estimate the performance gain. In this section, we will discuss two strategies to select the most promising indexed query patterns that to the most extent enhance query evaluation.

As far as we are concerned, all RDF repositories store and index RDF data in triples. The index to triples could be thought of as a pattern with two nodes and one edge or a node with constraint. A triple pattern ($?x$, EDGE_NAME, $?y$) corresponds to "*[+EDGE_NAME]*" in our approach and a triple pattern ($?x$, PROP, VALUE) with a property PROP of value VALUE is a pattern "PROP_NAME::VALUE". We use the materialized patterns above to guarantee the completeness of our index. Additional patterns are selected and added as auxilary source to enhance query perforamnce.

Two different indexing selection strategies will be presented in this paper. In the following subsections, we first show the motivation behind the two strategies and then describe them in detail.

5.1   Heuristic

This approach relies on the idea of trimming the pattern tree nodes that have large solution sets. By the definition of query pattern tree, if we select the parent

Algorithm 1. Heuristic Pattern Selection

    input: P : a set containing basic patterns; $maxsize$: maximum size of patterns;
         $threshold$: space limit.

1 $E \leftarrow P$;
2 $H \leftarrow$ heap built from P;
3 **while** H is not empty and $threshold$ is not reached **do**
4      $p \leftarrow$ pop H;
5   **if** $|V_p| < maxsize$ **then**
6   **forall the** $p' = p \diamond_x e$ not in E **do**
7   **if** cost reading $p' <$ lowest cost computing p' **then**
8                $E \leftarrow E \cup \{p \diamond_x e\}$;
9   Insert $p \diamond_x e$ into H.;
10   Store and index $p \diamond_x e$.;

11   **forall the** $p' = (V_p, E_p, c_p, h_{p'})$ of p not in E **do**
12   **if** cost reading $p' <$ lowest cost computing p' **then**
13            $E \leftarrow E \cup \{p'\}$;
14   Insert p' into H.;
15   Store and index p'.;

of a node as an indexed pattern, then the node itself will no longer appear in this tree as the division will terminate at its parent node. Inspired by this, we can index and store the super query patterns of those patterns with large solution sets. The effect of this algorithm is to trim off nodes of high cost from execution plans.

Our algorithm behaves as follows. Initially, all basic patterns are chosen as indexed patterns. The ones whose solution set sizes are above a threshold are inserted into a maximum heap ranked by the sizes of their solution sets. Iteratively, the pattern p on the top of the heap is popped out and gets extended to a new pattern p' either by adding an edge e connecting to node x, denoted as $p \diamond_x e$, or by imposing an extra hashing constraint, i.e. $p' = (V_p, E_p, c_p, h_{p'})$ where for certain node $u \in V_p$ with $h_p(u) = \varnothing$, $h_{p'}(u) = v$ and otherwise $h_{p'}(u) = h_p(u)$. If the extended p' is promising enough as an indexed pattern, its solution set will be calculated and stored, i.e. it would take less effort directly retrieving its solution set than calculating it from the current set of indexed patterns. p' would be inserted into the heap if the size of its matching set is larger than a threshold. The iteration of extension stops when the heap is empty or the space limit is reached. The pseudo code is shown in Algorithm 1.

5.2   Frequent Query Patterns

If the statistics on the query log are known in advance, we can make more accurate decision on pattern selection. We can spare more space for the patterns that are frequently visited than those that seldom are. We formalize the query log as the set $Q = \{q_i \mid q_i$ is a query input by a user$\}$. Given a query log Q,

Algorithm 2. Finding Frequent Query Pattern

> **input** : min_freq : minimum support of being a frequent pattern; Q : the given query log
> **output**: The set of frequent query patterns

```
1  E ← ∅; H ← empty heap;
2  forall the q ∈ Q do
3  │   forall the t ⪯p q with ht ≡ ∅ and ct ≡ ∅ and |Et| = 1 and |Vt| = 2 do
4  │   │   if support(t) ≥ min_freq and t ∉ E then
5  │   │   │   E ← E ∪ {t};
6  │   │   └   Insert t into H;

7  while H is not empty do
8  │   g ← pop H;
9  │   if |Vg| < maxsize then
10 │   │   forall the t = g ◇x e not in E with support(t) ≥ min_freq do
11 │   │   │   if cost reading t < lowest cost computing t then
12 │   │   │   │   E ← E ∪ {t};
13 │   │   └   └   Insert t into H;

14 │   forall the x ∈ Vg such that hg(x) = ∅ do
15 │   │   forall the super pattern t = (Vp, Ep, cp, ht) of p not in E with
   │   │   support(t) ≥ min_freq do
16 │   │   │   if cost reading t < lowest cost computing t then
17 │   │   │   │   E ← E ∪ {t};
18 │   │   └   └   Insert t into H;

19 return E
```

we define $support(g)$ as the percentage of graphs $q \in Q$ such that $g \preceq_p q$. The concept of frequent query patterns can be subsequently defined as graphs p with $support(p) \geq min\_freq$.

Algorithm 2 applies the pattern growth strategy as in gSpan [19] but it uses a Breadth First Search (BFS) strategy. The scope starting from line 2 is the initialization of BFS queue, where all frequent patterns containing only one edge are inserted. The BFS extension is performed in the scope beginning with line 7. The frequent pattern set grows either by adding an edge (in the scope starting from line 9) or by including a possible hash constraint (in the scope starting from line 14).

Assuming that we have generated a set of frequent query patterns, now comes a question that whether it is appropriate that we index all of them? Recall the presumption of Theorem. 1 that the cost of retrieval operator for an indexed pattern is always less than any other plans attempting to compute its matching set. We therefore sort all patterns according to the proportion of the costs of their retrieval operators to the costs of plans computing their matching sets and index from the most promising one until the space limit is reached.

Compared to the heuristic strategy, the frequent pattern-based strategy has a favor for for large and frequent patterns over small and rare ones. As smaller patterns would usually have larger solution sets, Different from the heuristic strategy preferring smaller patterns as those patterns would usually have larger solution sets, the frequent pattern based strategy would select bigger patterns, which may further increase the query performance. However, frequent pattern-based strategy is vulnerable to query log bias. If the query log is not representative of the queries inputted by the user, the performance would be diminished due to misses of some promising patterns.

6   Experiments

In our experiments, we compare the performance of pattern-based index using the two pattern selection strategies mentioned in Section 3 with a baseline. The baseline is an index which only stores the bindings of all the aforementioned basic triple patterns. Since the detailed implementation of the triple stores such as programming languages, transactions, and locks, also affects the execution time, our baseline uses the same execution operators under the same framework as the patter-based index to reduce these factors. For the baseline, we use the same dynamic programming planning algorithm and execute our queries following the 3 execution plans of lowest costs, among which the lowest execution time among the 3 plans is used as the baseline. This reflects a typical setting of a triple store such as Jena [17], Sesame [4], and RDF-3X [9] to which our approach also applies. But there are no histograms on join selectivities in our implementation as in RDF-3X but we choose the best performance among several optimal plans to approximate the best possible query performance under the triple index.

We opt for the inverted index and sequential fixed length records file in both triple-based and pattern-based approaches.

Our implementation has no page buffers so that every access to a new page is actually performed on disk. Moreover, they are non-transactional and no locks are needed before reading and writing. Two data sets are used in our experiments, DBPedia and LUBM[1]. DBPedia is a knowledge base extracted from WikiPedia and which has rich information on relations and properties. As an open knowledge base, it has no predefined schemas. LUBM is a synthetic data set, in which the types of relations and properties are relatively less than DBPedia. The two data sets' scale information is shown in Table 1.

Table 1. Dimensions of datasets

	DBPedia	LUBM50	LUBM100	LUBM150	LUBM200
#Nodes	2,414,030	1,066,943	2,148,715	3,197,823	2,785,068
#Edges	11,945,160	4,445,950	9,427,197	14,026,098	17,831,301
#Keywords	32,404,537	2,163,862	4,357,698	6,485,364	8,790,250

---

[1] http://swat.cse.lehigh.edu/projects/lubm/

Table 2. Index time(sec)

	Triple	Heuristic	Freq. Pt.
DBPedia	1413	2672	2514
LUBM50	224	641	1042
LUBM100	557	1759	2884
LUBM150	911	3218	5412
LUBM200	1285	5120	9133

Table 3. Disk usage(MB)

	Triple	Heuristic	Freq. Pt.
DBPedia	155	539	450
LUBM50	28	109	341
LUBM100	56	224	714
LUBM150	83	341	1085
LUBM200	113	450	1480

The experiments are conducted on a computer with an Athlon X2 5200+, 8GB RAM and a non-RAID SATA hard disk drive running Ubuntu Server 8.04 (64bit) operating system. The java environment is Sun JRE 1.6 (64bit) with a maximum 4GB memory.

6.1   Pattern-Based vs. Triple-Based

The performances of the Pattern-based and Triple-based approaches on two datasets (DBPedia and LUBM200) are compared and analyzed. For each data set, we generate query sets of different complexity, containing from one to three edges. $QS(i, j, k)$ represents a query set in which every query contains k edges and i nodes, j of which are labeled. Each query set contains 500 different queries. The probability of the occurrence of a query in a query set is proportional to the size of its matching set. In the heuristic strategy, a fixed amount of space limit is given. In the frequent query pattern strategy, we choose 20% of the queries in all query sets as our sampled query log and set min_freq to 0.3%.

In the DBPedia dataset, as shown in Fig. 8(a), our approach using heuristic strategy achieves at least double the efficiency of triple-based approach and using the frequent pattern strategy performs three to six times faster than the triple-based approach. In the LUBM dataset, as in Fig. 8(b), average query time using the heuristic strategy is reduced by three orders of magnitude compared to the traditional approach and using the frequent pattern strategy improves by one to two orders of magnitude. The causes of performance difference between the two strategies will be discussed later.

Table 2 and Table 3 have compared the index time and disk space used for each of the approaches being experimented. On both datasets our approach takes three to four times the disk space of the triple-based approach and around

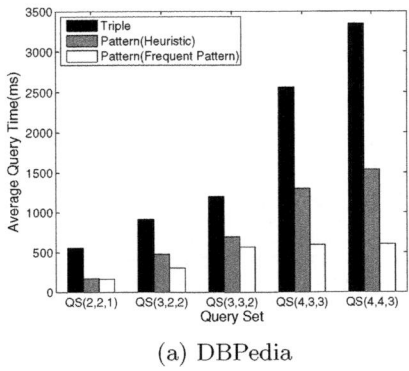
(a) DBPedia

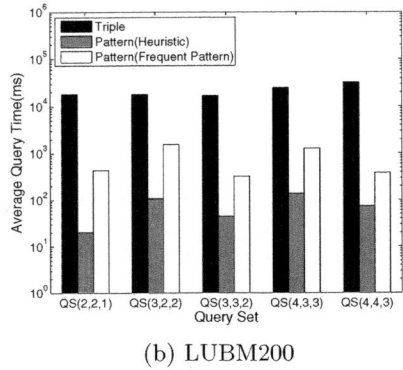
(b) LUBM200

Fig. 8. Comparison of query execution time

the same scale of increase with indexing time. The experiments show that our approach has enhanced query speed significantly while reasonably sacrificing the disk usage. The tradeoff between space and time will be discussed shortly.

6.2   Heuristic vs. Frequent Pattern

From Fig. 8(a) and (b), we see that the two strategies have diverse performance.

For the DBPedia dataset shown in Fig. 8(a), we find that two strategies' performances are close on simple query sets, e.g. $QS(2, 2, 1)$, $QS(3, 2, 2)$ and $QS(3, 3, 2)$. The gap widens in $QS(4, 3, 3)$ and $QS(4, 4, 3)$, where the frequent pattern strategy achieves roughly twice the efficiency of the heuristic one. This phenomenon reflects the advantage of frequent pattern strategy. With bigger patterns covering the query and fewer joins in plans, it achieves better performance.

On the contrary, as shown in Fig. 8(b), the heuristic strategy outperforms the frequent pattern one in LUBM. The reason attributed to the intrinsic characteristics of the data set. We observe that the matching sets of patterns with two constrained nodes are already very small. Indexing bigger patterns does little help to the evaluation speed. This is shown by the fact that the heuristic selection always stops at that point because all promising patterns are already indexed. However, frequent pattern strategy is biased as we have stated in Section 5.

6.3   Scalability

In order to examine the scalability of our approach, we adjust the scale factor of the LUBM dataset (in specific, the number of universities) from 50 to 200. The detailed sizes are shown in Table 1. Fig. 9 shows the average query time over all queries in a query set containing 1500 queries in different sizes. This figure demonstrates that the average query time of our approach is nearly linear to the scale of data, which is better than the quadratic of the triple-based approach.

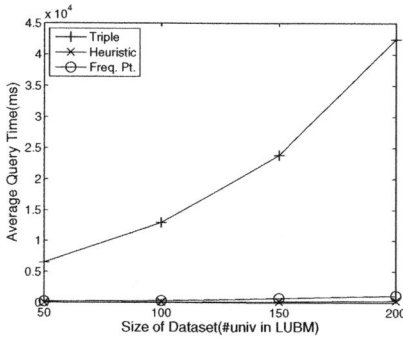

Fig. 9. Scalability

From Table 2 and Table 3, we see that the disk usage of our approach varies linearly with the scale of data and the index time is close to a linear function of the data scale, both conforming to the triple-based approach.

Based on the experimental results, we conclude that as long as the characteristics of data, specifically the average number of labels on each node, the number of edge types and the density of the data graph remains unchanged, our approach has the same or better scalability comparing to the triple-based approach.

6.4   Space-Time Trade-off

Fig. 10 depicts the relation between the average query time and the disk usage on the DBPedia dataset. We see that the average query time of both strategies monotonously decreases as the disk usage increases. Though selecting more patterns to index will enhance the performance, the trade-off may not be always worthwhile. The effect of incrementing indexed patterns will become less significant as we continuously select more patterns to index. This is because our indexing strategies preferentially select the more promising patterns and

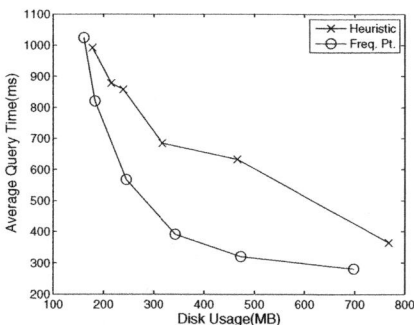

Fig. 10. Space-time tradeoff

maximize the performance gain. According to the curve shown in Fig. 8 and the criterion defined by the system designer, we can find the balance point between time and space.

7   Related Work

Publicized RDF stores use different techniques to enhance the query evaluation. Sesame [4], Jena [17] and Oracle[7] store triples in relational database tables. They rewrite SPARQL queries into SQL to leverage the query capability of relational DBMS. RDF-3X [9,10] and Hexastore [5] apply native implementation of index and query execution engine to speed up query performance. Hexastore permutates the subject, predicate and object resources to build 6 indexes. The triple patterns applied in Hexastore also can be indexed as patterns in our approach so that in a sense our approach generalizes Hexastore. RDF-3X shares a similar execution engine design with our approach but it takes no action to materialize joined results. It also relies on the join selectivity histogram to optimize join ordering. However, it still suffers from the performance problem of joining two large solution sets especially when they are not fit in the memory and require external merge join.

Sesame [4], Jena [17] and Oracle[7] use clustered-property table to speed up query execution but they require the application level design and tuning to define them. As a consequence, they are not suitable for data with the complex schema and flexible query pattern. Additionally, none of them reported performance studies on large scale datasets that contain millions of triples.

Vertical partitioning [1] was proposed to partition triples into narrow property tables with two columns and column-oriented databases [15] were employed as the query engine. It reported better performance over traditional RDF engines built on traditional relational databases but was outperformed by the native store RDF-3X. Sidirourgos et al. [13] also revisited the result and questioned vertical partitioning for scalability concerns. Another drawback of the solutions built on top of existing query engines is that the underlying physical design is a black box that is difficult to analyze, thus making further optimization impossible.

GRIN [16] divides triples into clusters and builds binary tree index from them. Their empirical study conducted on small size RDF data showed performance increase compared to RDF triple stores.

8   Conclusion

In this paper, we have proposed a pattern-based approach for storing and indexing RDF data, based on which an efficient method for processing SPARQL queries based on graph matching has been developed to take advantage of materialized pattern solutions. Two strategies based on heuristic rules and frequent patterns respectively, have been devised to select the best patterns included in indexes. Our experimental results show that the pattern-based approach significantly improves traditional approaches at the cost of a reasonable extra disk usage. The result also reveals the advantages and disadvantages of two introduced pattern selection strategies and further testifies that our approach achieves better scalability over traditional approaches.

Our future work includes extending our query capability to support full SPARQL queries and adapting current approach to column-oriented databases and distributed systems. By extending operators, we can further support complex SPARQL queries in addition to basic graph pattern queries. As a result, investigating possible patterns that can be used by these operators is also an important topic. On the other hand, column-oriented databases logically support sparse wide tables which well fit our need for the heap file. Substituting the index and storage with the column-oriented file structure and analyzing its performance are also planned. Following this line of research, our approach can be feasibly ported to distributed OLAP frameworks such as Hadoop and Hive. Materialized solutions will continue to better reduce the number of iterations and response time.

References

1. Abadi, D., Marcus, A., Madden, S., Hollenbach, K.: SW-Store: a vertically partitioned DBMS for Semantic Web data management. The VLDB Journal 18(2), 385–406 (2009)
2. Angles, R., Gutierrez, C.: The Expressive Power of SPARQL. In: Sheth, A.P., Staab, S., Dean, M., Paolucci, M., Maynard, D., Finin, T., Thirunarayan, K. (eds.) ISWC 2008. LNCS, vol. 5318, pp. 114–129. Springer, Heidelberg (2008)
3. Beckett, D.: RDF/XML Syntax Specification
4. Broekstra, J., Kampman, A., van Harmelen, F.: Sesame: A Generic Architecture for Storing and Querying RDF and RDF Schema. In: Horrocks, I., Hendler, J. (eds.) ISWC 2002. LNCS, vol. 2342, pp. 54–68. Springer, Heidelberg (2002)
5. Weiss, C., Karras, P., Bernstein, A.: Hexastore: sextuple indexing for semantic web data management. In: Proceedings of the VLDB (2008)
6. Chaudhuri, S.: An overview of query optimization in relational systems. In: Proceedings of the Seventeenth ACM SIGACT-SIGMOD-SIGART Symposium on Principles of Database Systems, PODS 1998, pp. 34–43 (1998)
7. Chong, E., Das, S., Eadon, G., Srinivasan, J.: An efficient SQL-based RDF querying scheme. In: Proceedings of the 31st International Conference on Very Large Data Bases, pp. 1216–1227. VLDB Endowment (2005)
8. Cook, S.: The complexity of theorem-proving procedures. In: Proceedings of the Third Annual ACM Symposium on Theory of Computing, pp. 151–158. ACM (1971)
9. Neumann, T., Weikum, G.: RDF-3X: a RISC-style engine for RDF. Proceedings of the VLDB Endowment 1(1), 647–659 (2008)
10. Neumann, T., Gerhard, W.: Scalable join processing on very large RDF graphs. In: Proceedings of the 35th SIGMOD, pp. 627–639 (2009)
11. Prud'hommeaux, E., Seaborne, A.: SPARQL Query Language for RDF
12. Schmidt, M., Meier, M., Lausen, G.: Foundations of SPARQL query optimization. In: Proceedings of the 13th International Conference on Database Theory, pp. 4–33. ACM (2010)

13. Sidirourgos, L., Goncalves, R., Kersten, M., Nes, N., Manegold, S.: Column-store support for RDF data management: not all swans are white. Proceedings of the VLDB Endowment 1(2), 1553–1563 (2008)
14. Silberschatz, A., Korth, H., Sudarshan, S.: Database system concepts, vol. 72. McGraw-Hill (2002)
15. Stonebraker, M., Abadi, D., Batkin, A., Chen, X., Cherniack, M., Ferreira, M., Lau, E., Lin, A., Madden, S., O'Neil, E., et al.: C-store: a column-oriented DBMS. In: Proceedings of the 31st International Conference on Very Large Data Bases, pp. 553–564. VLDB Endowment (2005)
16. Udrea, O., Pugliese, A., Subrahmanian, V.S.: GRIN: A graph based RDF index. In: Proceedings of the National Conference on Artificial Intelligence, vol. 22, p. 1465. AAAI Press, MIT Press, Menlo Park, Cambridge (1999/2007)
17. Wilkinson, K., Sayers, C., Kuno, H.: Efficient RDF storage and retrieval in Jena2. In: Proceedings of SWDB (2003)
18. Wilkinson, K.: Jena property table implementation. In: Proc. of the International Workshop on Scalable (November 2006)
19. Yan, X., Han, J.: gSpan: Graph-based substructure pattern mining. Order A Journal On The Theory Of Ordered Sets And Its Applications (2002)

The HɪLεX System
for Semantic Information Extraction

Marco Manna[1], Ermelinda Oro[2], Massimo Ruffolo[3], Mario Alviano[1],
and Nicola Leone[1]

[1] Department of Mathematics, University of Calabria, Italy
{manna,alviano,leone}@mat.unical.it
[2] DEIS, University of Calabria, Italy
oro@deis.unical.it
[3] ICAR-CNR, University of Calabria, Italy
ruffolo@icar.cnr.it

Abstract. The explosive growth and popularity of the Web has resulted
in a huge amount of digital information sources on the Internet. Unfortunately, such sources only manage data, rather than the knowledge they
carry. Recognizing, extracting, and structuring relevant information according to their semantics is a crucial task. Several approaches in the
field of Information Extraction (IE) have been proposed to support the
translation of semi-structured/unstructured documents into structured
data or knowledge. Most of them have a high precision but, since they are
mainly syntactic, they often have a low recall, are dependent on the document format, and ignore the semantics of information they extract. In
this paper, we describe a new approach for semantic information extraction that could represent the basis for automatically extracting highly
structured data from unstructured web sources without any undesirable
trade-off between precision and recall. In short, the approach (i) is ontology driven, (ii) is based on a unified representation of documents, (iii)
integrates existing IE techniques, (iv) implements *semantic regular expressions*, (v) has been implemented through Answer Set Programming,
(vi) is employed in real-world applications, and (vii) is having a positive
feedback from business customers.

1   Introduction

Context and Motivation. In the past decade, the number of digital information sources (such as document repositories, digital libraries and websites)
has risen exponentially. These sources represent a large amount of knowledge
for humans but not for machines. In fact, such knowledge is encoded mainly by
means of semi-structured and unstructured documents (hereafter non-structured
documents). Therefore, *identifying*, *extracting* and *storing* information from non-
structured documents are widely recognized as main issues in the field of *information and knowledge management* [32, 16, 22, 42, 37, 8, 39, 9, 23]. The issues of
identifying, extracting and storing information from non-structured documents
are referred to as Information Extraction (IE).

A. Hameurlain et al. (Eds.): TLDKS V, LNCS 7100, pp. 91–125, 2012.

Existing approaches to IE are mainly syntactic. Most of them use *regular expressions (regexes)*, a simple and declarative formalism for specifying patterns to be extracted. An advantage of regexes is that they are suitable for efficient evaluation. In fact, recognizing whether a string belongs to a regular language is feasible in linear time. However, a main limitation of regexes is their expressiveness, which is not sufficient for powerful IE tasks. For example, regexes cannot express patterns such as $a^n b^n$, representing strings made of n consecutive a's followed by n consecutive b's. Moreover, using regexes in large extraction tasks usually results verbose, chaotic and non modular.

Another shortcoming afflicting most of the existing approaches to IE is that they are strongly dependent on a specific document format. For example, systems conceived for handling html documents are usually not suitable for pdf or txt files. This limitation is mainly due to the pure syntactic approach adopted by these systems. Indeed, these approaches principally use pattern matching mechanisms (such as regexes) on textual fragments by possibly taking advantage of the underlying structure of documents (for example, in case of html or pdf files). Stated differently, within these systems, information is extracted regardless of its semantics.

Robust IE tasks should take advantage of the background knowledge of the domain of interest. In particular, the background knowledge of many relevant domains (such as medical or linguistic) is already represented in the form of *ontologies*, one of the most commonly accepted formalism for knowledge representation. Moreover, robust IE tasks require high-level patterns that cannot be expressed by pure syntactic approaches, such as those based on regexes. Patterns of this kind are expressible, for instance, by formal grammars endowed with attributes, referred to as *attribute grammars*. The attribute grammar formalism is a declarative language introduced by Knuth as a mechanism for specifying the semantics of context free languages [36]. Therefore, attribute grammars seems to be a natural and declarative formalism for describing object-oriented patterns for semantic information extraction. However, even if attribute grammars allow for specifying *semantic equations*, they cannot be considered a knowledge representation formalism per se. In fact, as previously observed, a semantic approach to IE should take advantage of domain knowledge. Our objective is thus to define a powerful and high-level extraction language by combining attribute grammars and ontologies. The resulting language is suitable for knowledge representation and extraction.

Contribution. The main contributions of this paper are as follows:

– We define a novel extraction language for specifying high-level and object-oriented semantic rules. These rules are *semantic regular expressions* used for discovering and extracting objects from non-structured documents.
– We discuss complexity issues regarding the proposed language. In particular, even if general attribute grammars require exponential time in the worst case [15], we identify a relevant tractable class by imposing non severe restrictions.

– We design and implement H*ıL*εX, an advanced system for ontology-based
  IE, that is founded on the aforementioned language and has already received positive feedback from business customers.[1] We give a panorama of its
  architecture and survey some real-world applications.

Our approach has the following features:

– It integrates existing techniques such as pattern matching, pattern recognition, natural language processing, machine learning, and heuristics.
– It is based on a unified 2-dimensional representation of non-structured documents, that is, documents are partitioned into adjacent sub-areas, each of
  which consists of a single text line.
– It is *ontology driven* in the sense that it uses *ontologies* for representing both
  input documents and the knowledge characterizing a specific domain. More
  specifically, ontologies are represented in OntoDLP [55], an object-oriented
  language based on Answer Set Programming (ASP).
– It implements *semantic regular expressions* for specifying high-level and
  object-oriented semantic rules for discovering and extracting information
  from different types of documents. For example, the following patterns have
  natural representations in our language:

  • *select* a string w followed by any member of class A;
  • *select* a member of class A related to a member of class B;
  • *create* a member of class A if other *selected* objects satisfy a given
    pattern.

Related Work. Programs that perform IE tasks are referred to as *extractors*
or *wrappers*. More specifically, wrappers locate relevant patterns of data items
on web pages, extract them, and transform them into structured data.

Several academic approaches have been proposed, and a number of commercial
enterprises supply the market today. Examples of relevant academic approaches
includes Araneus [47], DEByE [53, 41], Minerva [11], NoDoSE [1], RAPIER
[6], SoftMealy [32], Squirrel [7], SRV [24, 25], STALKER [48, 49], TSIMMIS
[30, 29], W4F [57], WebOQL [3], WHISK [58], WIEN [40, 38], and XWrap [45].
Some commercial enterprises developing IE products are Denodo, Fetch Technologies, Kapow Technologies, Lixto, and QL2. These academic and commercial
approaches mainly provide wrappers that are either programmed manually, or
learned by semi-automatically supervised systems while a user interacts with
example documents. We remark that a main shortcoming of these approaches is
their lack of understanding of extracted information.

Some approaches that move towards fully automatic or generic wrapping of
the existing World Wide Web have been proposed and are currently under active development — see for instance DIPRE [5], KnowItAll [21], MDR [44],

---

[1] This feedback is confirmed by the industrialization of the system by Exeura Srl, a
technology company working on analytics, data mining, and knowledge management.

RoadRunner [13, 12], Snowball [2], TextRunner [4]. However, although they show significant progress, they still lack the combined recall and precision necessary to allow for very robust queries.

Several works have also shown the promise of deducing and encoding formal knowledge in the form of ontologies. These approaches use ontologies either to improve the extraction phase as a way to present the results of the extraction, or to allow matching different representations across sources — see for instance PANKOW [10], TANGO [18, 19, 17] TARTAR [50], KYLIN [59].

For a more in depth classification of the aforementioned wrappers we refer to [32, 42, 37, 8, 39], where the following aspects are considered: type of input documents (e.g., html, pdf, txt); page contents (e.g., columns, tables, figures); extraction patterns or methods (e.g., regular expressions, Prolog-like logic rules, NLP rules, heuristics, probabilistic hidden Markov models); support for complex objects; degree of automation (i.e., manual, semi-supervised or automatic systems); background knowledge (i.e., annotation examples, taxonomies, ontologies); output (e.g., databases, XML output, ontologies).

Organization. The remaining of the paper is organized as follows. Section 2 introduces the reader to the concept of an ontology, formally defines both HιLεX ontologies and query answering over such ontologies, and gives an informal overview of the OntoDLP language. Section 3 discusses issues regarding the ontological representation of documents. Section 4, after a brief overview of the language, presents syntax and semantics of *semantic regular expressions*, also referred to as *descriptors*. Section 5 considers linguistic restrictions on descriptors ensuring tractability, as well as complexity issues. Sections 6 presents the general architecture of the HιLεX system, and briefly surveys some of its successful applications. Section 7 draws our conclusions.

2   Ontologies

In Artificial Intelligence, different informal definitions of the term *ontology* have appeared in the literature [26–28]. For example, in [26]: "An ontology is a specification of a shared conceptualization of a domain." From a formal point of view [14, 52], an ontology is a quadruple $\langle C, R, \iota, \xi \rangle$, where C is a set of concepts, R is a set of relations, ι is a set of instances, and ξ is a set of axioms. By following and specializing this definition, we now give the notion of ontology used in HιLεX.

2.1   HιLεX Ontologies

Let Z be a fixed, countable set of *constants*. Let $\tilde{Z}$ denote the set of *values* obtained by the union of Z with the set of finite lists of elements from Z (including the empty list). We formally define an *ontology* on Z as an eight-tuple $\mathcal{O} = \langle D, A, C, R, \preceq, \sigma, \delta, \iota \rangle$ where:

- D, A, C, R are finite, disjoint sets of entity names respectively called *datatypes*, *attribute-names*, *classes* and *relations*. The set A contains the special attribute-name oid. Elements in $C \cup D$ are called *flat-types*. For each flat-type t, there is a *list-type* denoted by $[t]$. Their union (flat-types and list-types) forms the set of all *types* denoted by T.

- $\preceq$ is a partial order (called *isA*) on C.
- $\sigma : C \cup R \to 2^{A \times T}$ is the *schema* function. Given an element e in $C \cup R$, the set $\sigma(e)$ is *the schema of* e. Any pair (a, t) in $\sigma(e)$ is the schema-*attribute* of e with name a and type t. The schema of a class c contains the attribute (oid, c), whereas no relation schema contains attributes with name oid.
- $\delta : D \to 2^Z$ (the *domain* function) associates a *value domain* with each data-type.
- ι is the *instance* function associating to each class or relation a possibly empty set of elements from $2^{A \times \tilde{Z}}$. Let $e \in C \cup R$ and $\hat{\imath}$ be an element of $\iota(e)$. Such an element $\hat{\imath}$ is called *instance of* e. Each pair (a, z) in $\hat{\imath}$ is the instance-*attribute* of $\hat{\imath}$ with name a and value z. If e is a class, then $\hat{\imath}$ contains exactly one pair of the form (oid, z_o). The value z_o is the object identifier (*oid*, for short) of $\hat{\imath}$. Class instances and relation instances are also called *objects* and *tuples*, respectively.

In addition, let t_1, t_2 be two types. We say that t_2 is a *subtype of* t_1 if at least one of the following conditions is true:

- $t_1 = t_2$;
- t_1, t_2 are classes and $t_2 \preceq t_1$;
- $t_1 = [\hat{t}_1], t_2 = [\hat{t}_2]$ are list-types and $\hat{t}_2$ subtype of $\hat{t}_1$.

Moreover, let t be a type. A value z is *compatible* with t if at least one of the following conditions is true:

- t is a data-type and $z \in \delta(t)$;
- t is a class and z is an *oid* of an instance of t;
- $t = [\hat{t}]$ is a list-type, z is a list, and each value in z is compatible with $\hat{t}$.

Finally, $\mathcal{O}$ must satisfy the following consistency restrictions:

- No class schema nor relation schema has two attributes with the same name.
- Let c_1, c_2 be two classes such that c_2 is a subtype of c_1. For each attribute (a, t_1) in the schema of c_1 there is one attribute (a, t_2) in the schema of c_2 such that t_2 is a subtype of t_1.
- Let e be either a class or a relation. Each instance of e has cardinality $|\sigma(e)|$.
- Let e be either class or a relation, and $\hat{\imath}$ be an instance of e. A pair (a, z) belongs to $\hat{\imath}$ only if both (a, t) is in $\sigma(e)$ and z is compatible with t.
- Let c_1, c_2 be two classes such that c_2 is a subtype of c_1. Let $\hat{\imath}_1, \hat{\imath}_2$ be two instances of c_1, c_2, respectively. Thus, $\hat{\imath}_1 \subseteq \hat{\imath}_2$.
- Let c_1, c_2 be two classes such that c_2 is a subtype of c_1. Let $\hat{\imath}_2$ be an instance of c_2, and z be its *oid*. There is precisely one instance of c_1 having z as *oid*.
- Let c_1, c_2 be two classes incomparable under *isA* ($c_1 \not\preceq c_2$ and $c_2 \not\preceq c_1$). No pair of instances of c_1 and c_2 can share the same *oid*.

2.2   Query Answering

We now introduce the notion of query and query answering over ontologies. Let $V = \{x_1, x_2, \ldots\}$ be a fixed, countable set of *variables* which, intuitively, will take values from the universe $\tilde{Z}$. A *term* is either a value in $\tilde{Z}$ or a variable in V. An *atom*, say m, is an expression of the form $e(a_1 : w_1, \ldots, a_q : w_q)$ such that: e is an element in $C \cup R$; a_j is an attribute-name in A; and w_j is a term for each $j \in \{1, \ldots, q\}$. If e is a class, then m is a *class-atom*, otherwise m is a *relation-atom*. Atom m is ground if w_j is from $\tilde{Z}$. Atom m is *true* with respect to $\mathcal{O}$ if there exists an instance $\hat{\imath} \in \iota(e)$ such that each pair (a_j, w_j) belongs to $\hat{\imath}$.

A *query* $\mathcal{Q}$ on $\mathcal{O}$ is a conjunction $m_1, \ldots, m_n$ of atoms. A *substitution* for $\mathcal{Q}$ is a mapping μ from the variables occurring in $\mathcal{Q}$ to values in $\tilde{Z}$. Let $\mu(\mathcal{Q})$ denote the ground query obtained from $\mathcal{Q}$ by replacing each variable x by $\mu(x)$. The set of all substitutions μ such that all ground atoms of $\mu(\mathcal{Q})$ are true with respect to $\mathcal{O}$ is denoted by $ans_{\mathcal{Q}}(\mathcal{O})$.

2.3   The OntoDLP Language

In this section, we describe the OntoDLP language [55] implementing H*ı*L*ε*X ontologies. The presentation of OntoDLP is given by examples and structured in paragraphs describing the most relevant constructs of the language (i.e., classes, objects, lists, taxonomies, relations, and queries).

Classes. A *class* can be thought of as a collection of individuals who belong together because they share some properties. Classes can be defined in OntoDLP by using the keyword class followed by its name. Class attributes can be specified by means of pairs *(attribute-name : attribute-type)*, where *attribute-name* is the name of the property and *attribute-type* is the class (or data-type such as int, string) the attribute belongs to. Suppose the aim is to model the domain of a *banking* enterprise, and some classes of individuals have been identified, namely: *banks, branches, accounts, persons, enterprises* and *places*. Suppose also that a number of relevant properties (or attributes) shared by all the individuals belonging to these classes are recognized. For instance, it is known that: banks have a name and own an asset; the branches of a given bank are located into a given place and also have an asset; accounts have a balance; enterprises have a name and a country (which is a place); persons have name, age, residence (which is also a place), father and mother (which are other persons); and finally, each place has a name. The above-listed classes and (related) attributes can be represented in OntoDLP by the following class-schemas:

```
class bank(name:string, asset:int).
class account(balance:int).
class place(name:string).
class branch(partOf:bank, location:place, asset:int).
class enterprise(name:string, country:place).
class person(name:string, age:int, father:person,
             mother:person, residence:place).
```

Note that class attributes in OntoDLP model the properties that *must* be present in all class instances; *optional* properties should be modeled by using other constructs such as lists, and relations, that will be described later in this section.

Objects. Domains contain individuals which are called *objects* or *instances*. Each individual in OntoDLP belongs to a class and is uniquely identified by a constant called *object identifier* (*oid*). Objects are declared by asserting a special kind of logic facts (asserting that a given instance belongs to a class). For example, with the facts

```
rome:place(name:"Rome").
john:person(name:"John", age:34, father:jack, mother:ann, residence:rome).
```

it is declared that *rome* and *john* are instances of `place` and `person`, respectively. Note that, when an instance is declared, an *oid* is immediately given to the instance (e.g., `rome` identifies a place named `"Rome"`), which may be used to fill an attribute of another object. In the example above, attribute `residence` is filled with the *oid* `rome` modeling the fact that *john* lives in Rome; in the same way, `jack` and `ann` are suitable *oids* respectively filling the attributes `father` and `mother` (both of type person).

Lists. A feature of OntoDLP is the possibility of exploiting as type a *list* (ordered collection) of instances that accepts *multiple copies* of the same instance. Given a class C, one can define the class *list of* C, denoted by $[C]$, having as instances all lists of individuals belonging to class C. For instance, the class `[string]` represents the class having as instances all lists of strings. Analogously, [*"This"*, *"That"*] is the list containing the string *"This"* followed by *"That"*. Lists are very useful for representing multi-valued attributes. Suppose that in our *banking* domain an account has an associated set of services that can be bought from the bank, e.g., an internet banking access, an overdraft protection, a payroll payment, etc. Suppose also that account services are represented as follows:

```
class accountService(cost:int).
   internet:accountService(cost:10).
   payroll:accountService(cost:1).
   ...
```

The definition of class `account` is upgraded by adding a new attribute `services` of type "list of `accountService`":

```
class account(services:[accountService], balance:int).
   a0001:account(services:[internet,payroll], balance:2000).
```

where the second one is an instance of `account` having associated the two services internet banking and payroll payment. (Note that [ ] denotes the empty list.)

Taxonomies. Concepts in an ontology are usually organized in taxonomies by using the *specialization/generalization* mechanism (which is called *inheritance* in object-oriented languages). This is done when a subset of individuals has some attributes that are not shared by all other individuals in the same class. For instance, employees are a special category of persons having extra attributes, like `salary` and `company`. OntoDLP supports inheritance by means of the special binary relation `isa`. In particular, the above-mentioned employee class can be declared as follows:

```
class employee isa person(salary:int, company:enterprise).
```

In this case, `person` is a more generic concept or *superclass* and `employee` is a specialization (or *subclass*) of `person`. Moreover, an instance of `employee` will have the local attributes `salary` and `company`, other than `name`, `age`, `father`, `mother` and `residence`, which are defined in `person`. We say that the last set of attributes is *inherited* from the superclass `person`. Hence, each proper instance of `employee` will also be automatically considered as an instance of `person` (the opposite does not hold). For example, the instance:

```
bob:employee(name:"Robert", age:25, father:jack, mother:betty,
             residence:rome, salary:2000, company:microsoft).
```

is automatically considered as an instance of `person` as follows:

```
bob:person(name:"Robert", age:25, father:jack,
           mother:betty, residence:rome).
```

Note that it is not necessary to assert the latter instance. Inheritance can be further applied to refine the design of our *banking* ontology. For instance, banks usually offer two different kinds of accounts, `checking` and `savings`. Moreover, inheritance can be applied repeatedly, without limitation on the number of superclasses (i.e., multiple inheritance is allowed). For instance, a bank may offer two special types of checking account: *gold account* having a fixed minimum balance; and *young account*, which is reserved to customers aged up to 21 years, and is, at the same time, both a saving account and a checking account. This may be specified in OntoDLP as follows:

```
class checkingAccount isa account(overdraftAmount:int).
class savingsAccount isa account(interestRate:int).
class goldAccount isa checkingAccount(minimumBalance:int).
class youngAccount isa savingsAccount, checkingAccountg.
```

Note that all list types are part of the OntoDLP inheritance hierarchy. Basically, the class [A] is a *subtype* of class [B] if and only if A is a subclass of B. For example, the class [savingsAccount] is subtype of [account] since savingsAccount is subclass of account. Finally, OntoDLP has a common built-in superclass called object which, apart from data-types (such as int and string), is superclass of all the classes defined by users.

Relations. Another important feature of an ontology language is the ability to model relationships among individuals. Relations are declared like classes: the keyword `relation` (instead of `class`) precedes a list of attributes. The set of attributes of a relation is called *schema* as for classes, and the cardinality of the schema is called *arity*. As an example, we model a relationship between persons and their bank account as follows:

```
relation customerHoldsAccount(customer:person, holds:account).
```

The instances of a relation are called *tuples*, and they can be declared by using logic facts. For instance, it can be asserted that `john` holds account `acc001` by writing a logic fact. Moreover, (since an account may be held by one more customer) two facts can be specified that assign the ownership of account `acc012` to both `ann` and `john`:

```
customerHoldsAccount(customer:john, holds:acc001).
customerHoldsAccount(customer:ann, holds:acc012).
customerHoldsAccount(customer:john, holds:acc012).
```

Contrary to class instances, tuples are not equipped with an *oid*.

Queries. Knowledge implicitly contained in an ontology can be retrieved by means of queries, as described in Section 2.2. As an example, the list of persons having a father who lives in Rome is requested as follows:

```
X:person(father:F), F:person(residence:P), P:place(name:"Rome")?
```

Note that attributes of atoms can be omitted when they are not of interest for the query.

3   Ontology-Based Document Representation

In this section, we describe the strategies that we have employed for decomposing a document, mainly *tokenization* and *part-of-speech tagging*. For each of these strategies, we also provide an ontological representation of extracted information. Moreover we briefly describe other annotation strategies as well as an ontological representation of raw strings. Our analysis is focused on a single text line, even though the HιLεX system uses a 2-dimensional document representation (see Section 4.1 for details). In fact, a document (e.g., txt, pdf, html) is partitioned into adjacent sub-areas (see [35, 33, 61, 60, 51, 20, 31, 50] for successful approaches), each of which consists of a single text line. However, the ontological representation of 2-dimensional objects mimics the one for 1-dimensional objects.

Tokenization. Consider the classical *tokenization process* on strings,[2] in which whitespaces are ignored. The following ontology structures are used for representing the tokens of a string:

---

[2] Tokenization is the process of mapping sentences from character strings into strings of words.

```
class object.
class token isa object(value:string).
relation position(obj:object, start:int, end:int).
```

In particular, class `object` is assumed to be the highest superclass (i.e., root class) of the ontology. Moreover, class `token` stores the smallest textual units composing the original string by means of the attribute `value`. Relation `position` locates each token (as well as any other possible object) by the attribute `start` and `end` storing the position of its first and last character, respectively. Finally, `string` and `int` are data-types. For example, consider the following witness string:

I must close the door since the thief is very close.

The following set of instances are produced by tokenization:

```
tk01:token(value:"I").        position(obj:tk01, start:0, end:1).
tk02:token(value:"must").     position(obj:tk02, start:1, end:5).
tk03:token(value:"close").    position(obj:tk03, start:5, end:10).
            ⋮                              ⋮
tk10:token(value:"very").     position(obj:tk10, start:32, end:36).
tk11:token(value:"close").    position(obj:tk11, start:36, end:41).
tk12:token(value:".").        position(obj:tk12, start:41, end:42).
```

Note that tokens are considered consecutive since whitespaces are ignored. Note also that the positional representation used by the system is completely hidden to the user. Moreover, we use a unique *oid* for each token because we may associate different properties to the same char-sequence recognized as a token but appearing in different parts of the string.

Part-of-speech Tagging. A more sophisticated strategy for decomposing a string, mainly implemented by Natural Language Processing tools (NLP-tools), is the *part-of-speech tagging*. This strategy consists of marking up the words in a text as corresponding to a particular part of speech (nouns, verbs, adjectives, adverbs, etc.). As an example, consider to have the following ontology structures:

```
class word isa object(content:string, lemma:string, postag:string).
relation syn(lemma1:string, lemma2:string, postag:string).
        syn(lemma1:"close", lemma2:"near", postag:"adj").
    ⋮
```

where class `word` stores the *content words* composing the original string, each of which is associated with its *lemma*[3] and its *part-of-speech tag*, and relation `syn` collects synonymous lemmas according to their word-category. In this case, an NLP tool processing our witness string would produce the following instances of `word` and `position`:

---

[3] A lemma is a canonical word representing a set of words with the same meaning.

```
w01:word(content:"I", lemma:"I", postag:"pns").
w02:word(content:"must", lemma:"must", postag:"verb").
w03:word(content:"close", lemma:"close", postag:"verb").
        ⋮
w09:word(content:"is", lemma:"be", postag:"verb").
w10:word(content:"very", lemma:"very", postag:"av").
w11:word(content:"close", lemma:"close", postag:"adj").

position(obj:w01, start:0, end:1).
position(obj:w02, start:1, end:5).
position(obj:w03, start:5, end:10).
        ⋮
position(obj:w09, start:30, end:32).
position(obj:w10, start:32, end:36).
position(obj:w11, start:36, end:41).
```

where, for instance, the part-of-speech tag `pns` stands for "personal pronoun". Note that the lemma of a word may be different from its content (this is the case of word w09), or the content of a word instance can be composed by more than one token viewed as a unique lemma (e.g., *pop-up*). Finally, each text portion, appearing in different parts of the original string and having different properties, is associated with different objects (see token `"close"` in the example).

Other Annotation Strategies. The system can also make use of other techniques. For example, we can represent each *symbol* of the input as an instance of the class `symbol`. This, however, implies having to explicitly include and manage whitespaces in all the other decomposition strategies. Moreover, we can recognize regular structures such as *email addresses* or *dates*. Class schemas for these three decompositions are:

```
class symbol isa object(value:string).
class email isa object(user:string, domain:string).
class date isa object(d:int, m:int, y:int).
```

The following are the recommendations to be taken into account for producing valid decompositions:

1. Each decomposition process has to generate *non-overlapping* text units, namely it has to:
   - create a partition of the input string;
   - choice exactly one subset of this partition; and
   - give an ontological representation of this subset.
2. Provide at least one decomposition representing an exact partition of the input string. More specifically, such a decomposition has to produce consecutive text units, by possibly ignoring special symbols. For example, the proposed tokenization or the symbol decomposition comply with this requirement.

3. Provide at most one decomposition per class.
4. Any class with an associated decomposition must be direct subclass of `object`.
5. Include whitespaces either in all decompositions or none.
6. Decompositions should create objects that can be concatenated to the objects produced by the exact partitions. For example, with respect to our witness string, we should avoid to produce an object for the char-sequence *"nce the thief is ve"* since it cannot be concatenated to the other tokens.

Note that recommendation 6 is desirable but not mandatory.

Raw Strings. The HıLɛX system also supports an ontological representation of raw strings by means of the following relation schema:

```
relation str_pos(str:string, start:int, end:int).
```

which is similar to relation schema `position`. Since an input line of length n contains $\mathcal{O}(n^2)$ substrings, materializing all of the tuples of relation `str_pos` would affect the system performance. For this reason, the HıLɛX system only generates strings used in descriptors (see Section 4.2). We point out that the raw string representation is not an input decomposition, and thus has not to comply with the recommendations for valid decompositions.

4   The HıLɛX Language

In this section, we define the *semantic rules*, also called (object) *descriptors* for short, used to discover and extract objects from non-structured documents. We first describe the language of the HıLɛX system by examples, providing the intuitive meaning of the main constructs. After that, we formally give its syntax and its semantics. Linguistic restrictions and complexity issues are give in Section 5.

4.1   An Overview

Consider the web page shown in Figure 1(a) and the associated decomposition shown in Figure 1(b). Such a decomposition can be obtained by means of a *table recognition tool* implementing techniques recently described in the literature [35, 33, 61, 60, 51, 20, 31, 50] and which are out of the scope of this paper. In particular, the content of each cell is considered as a unique text line as described in the previous section, even if the content of the cell spans multiple lines in the document.

Example 1. The simplest and most basic rules are queries on the document working as in classical Web search engines. For instance, we may *select* the occurrences of the text fragment *"Mostly Sunny/Wind"* by only writing:

```
<"Mostly Sunny/Wind">   or   <"Mostly"> <"Sunny/Wind">
```

where the first query matches an exact char-sequence, while the second query matches a *concatenation* of two char-sequences possibly ignoring whitespace separators.   □

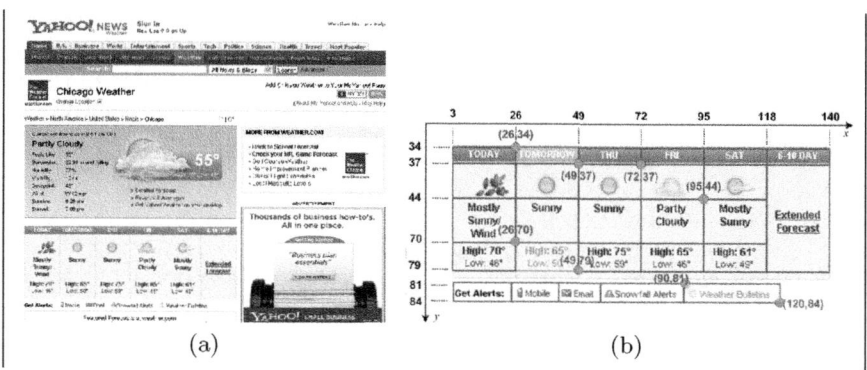

Fig. 1. Document Decomposition

Example 2. We may *select* the occurrences of both *"Sunny"* and *"Partly Cloudy"* by means of the following rule:

```
<"Sunny"> ∨ <"Partly Cloudy">
```

where the logical operator ∨ means *one of*, or we may *select* the cells (or boxes) containing occurrences of both *"High"* and *"Low"* by:

```
<box> CONTAIN <"High"> ∧ <"Low">
```

where the logical operator ∧ means *all of*. Notice that in this case we are limiting the scope of the query to objects of the class box. Clearly, such a scope might be in general any object, even the entire document.   □

Example 3. We may *select* any token (such as, *"FRI"*, *"TODAY"*, *"Sunny"*, *"61"*, *"Mobile"*) of the document by only writing:

```
<token>
```

or we may *select* tokens of length three (such as, *"FRI"*, *"Low"*) by:

```
<T:token(value:V), #hasLength(V,3)>
```

where T is a variable representing a concrete token, V is a variable representing the string value of T, and #hasLength is a built-in boolean function checking whether the first attribute is a string of length equals to the value of the second attribute. Note that variable T could be omitted in this example since it is not involved in any join; in this case, the system would return the result by means of an anonymous variable. Conversely, variable V cannot be omitted since it is involved in a join.   □

Example 4. We might *select* any word (such as *"Cloudy"*) whose lemma is synonymous of the adjective *overcast*:

```
<word(lemma:L), syn(lemma1:L, lemma2:"overcast")>
```

where syn and word are defined in Section 3.   □

Example 5. Atomic queries can be combined to recognize objects already present in the ontology. Let us assume that the ontology contains a class `wDay` (day of the week) with exactly seven instances, `wd1 ... wd7`, one for each day of the week. We may recognize a *day*, for example *Monday*, by means of the following rules:

```
<wd1> :: <token(value:"Monday")>.
<wd1> :: <token(value:"MON")>.
<wd1> :: <token(value:"Today"), #now("Monday")>.
<wd1> :: <token(value:"Tomorrow"), next(WD,wd1), WD:wDay(name:N), #now(N)>.
```

where the built-in unary boolean function `#now` returns true if and only if the running-day is the day specified by the attribute, and the binary relation `next` gives a total order between the days of the week. Having these descriptors, we may *select* strings referring to some day of the week (such as *"TODAY"*, *"THU"*, *"SAT"*) by writing:

```
<wDay>   or   <wDay(name:"Friday")>   or   <wd5>
```

where the first query selects all strings referring to some day, while the other two queries are equivalent and select strings referring to the fifth day of the week, that is, Friday.   □

Example 6. The system allows for creating new objects. For example, assume that we want to *select* a portion of text containing a moderate temperature variation between night and day, namely being at most 15 degrees. We define a class `moderateVar` having two integer attributes for keeping the two temperatures, and the following rule:

```
<moderateVar(I1,I2)> :: <"High:"> <token(value:V1), #digits(V1)>
   {I1:=$toInt(V1);} <token>* <"Low:">
   <token(value:V2), #digits(V2), I1-$toInt(V2)≤15> {I2:=$toInt(V2);}.
```

where `#digits` is a unary boolean function checking whether its parameter is a string representing an integer, and `$toInt` converts a string into an integer. Instructions in curly brackets set the values of the attributes. The application of this rule on the document shown in Figure 1 would produce two objects associated to the areas containing the strings "High: 65° Low: 50°" and "High: 61° Low: 49°". In fact, this rule can be intuitively read as follows: If there is a string "`High:`" followed by a token of only digits, then store in variable `I1` its decimal value. Now, ignore any token until there is a string "`Low:`". If this string is followed by a token of only digits, then compute its decimal value, compute the variation, and check whether it is at most 15. Thus, store in variable `I2` the decimal value of the second temperature. Finally, if everything has been successful, then create a new instance of the class `moderateVar` having `I1` and `I2` as attribute values.   □

Example 7. Assume that our ontology contains a binary, symmetric relation `famousCouple` between instances of class `person`, and that some rule has been already defined to recognize names and surnames of people. If we want to *select* text fragments referring to people connected through relation `famousCouple`, then we can define a class `gossip` with two attributes of type `person` and write the following rule:

```
<gossip(P1,P2)> :: <P1:person> <T:token, not sentenceTerminator(T)>[0..3]
    <P2:person, famousCouple(P1,P2)>
```

where `sentenceTerminator` is a unary relation populated by an NLP tool selecting tokens recognized as string terminators. The directive `[0..3]` means a (possibly missing) sequence of at most three tokens.   □

4.2   Syntax

Given an ontology over the a set Z set of values, we assume that Z always contains two subsets Z_{int} and Z_{str} standing for the integer numbers and all the strings over a given alphabet (or charset), respectively. Let V be a set of variables. First of all, we define all the expressions allowed in a descriptor. Afterwards, we formally introduce its whole structure.

Arithmetic Expressions. An *arithmetic expression* over $\{|\ |, +, -, \div, \%, *\}^4$ is inductively defined as follows:

- Any term $x \in V \cup Z_{int}$ is an arithmetic expression;
- If E is an arithmetic expression, then also $|E|$ is;
- If E_1, E_2 are arithmetic expressions, then also $(E_1 + E_2)$, $(E_1 - E_2)$, $(E_1 \div E_2)$, $(E_1 \% E_2)$, and $(E_1 * E_2)$ are.

String Expressions. A *string expression* over $\{\oplus\}$ is inductively defined as follows:

- Any term $x \in V \cup Z_{str}$ is a string expression;
- If E_1, E_2 are string expressions, then also $E_1 \oplus E_2$ is.

List Expressions. A *list expression* over $\{\oplus\}$ is inductively defined as follows:

- Any term $w \in V \cup (\tilde{Z} \setminus Z)$ is a list expression;
- If $w_1, \ldots, w_q$ are terms in $Z \cup V$, then $[w_1, \ldots, w_q]$ is a list expression;
- If E_1, E_2 are list expressions, then also $E_1 \oplus E_2$ is.

Comparison Predicates. Let E_1 and E_2 be either two arithmetic expressions or two string expressions. A *comparison predicate* over $\{<, \leq, =, \neq\}$ is one of the following binary boolean functions $(E_1 < E_2)$, $(E_1 \leq E_2)$, $(E_1 = E_2)$, and $(E_1 \neq E_2)$. Although their meaning should be clear for integers, we want to better explain what they mean for strings: E_1 is a strict substring of E_2, E_1 is substring of E_2, E_1 is exactly the same to E_2, and finally E_1 and E_2 are incomparable (none of $E_1 < E_2$, $E_2 < E_1$, $E_1 = E_2$ holds), respectively.

---

[4] The symbols in the set $\{|\ |, +, -, \div, \%, *\}$ denote the operators *absolute value, addition, subtraction, integer division* with truncation, *modulo reduction* (or remainder from integer division), and *multiplication*, respectively. Note that we only consider the integer division, and thus any arithmetic expression over these operators returns an integer value.

Boolean Formulae. A (boolean) *formula* over $\{\wedge, \vee, \neg\}$[5] is inductively defined as follows:

- Any comparison predicate P is a formula;
- If P is a formula, then also $\neg P$ is;
- If P_1 and P_2 are formulae, then also $(P_1 \wedge P_2)$, and $(P_1 \vee P_2)$ are formulae.

Instructions. An *instruction* is one of the following:

- [*type_name*] *variable_name* ':=' *value* ';'
- [*type_name*] *variable_name* ':=' *arithmetic_expression* ';'
- [*type_name*] *variable_name* ':=' *string_expression* ';'
- [*type_name*] *variable_name* ':=' *list_expression* ';'
- [*type_name*] *variable_name* ':=' *formula* '?' *expr_1* ':' *expr_2* ';'

where the *type_name* appears only in instructions declaring a new variable.

Atomic Body Elements. An *atomic body element* is one of:

- a *query* starting with a *class-atom*;
- an *oid*;
- a value from Z_{str};

possibly followed by a sequence of instructions, with the restriction that each variable can be assigned at most once.

Actually, with no rise in complexity, to add more constraints among the values of the variables, we also allow boolean formulae in conjunction with the atoms of a query. Examples of atomic body elements are:

```
<"a"> {int N:=1;}
<w01> {int N:=N+1;}
<adjective(word:W), syn(W,"small")> {Str:=Str⊕","⊕W; N:=N+1;}
<P:person(numChildren:NC), NC=2*Nmin ∧ NC≠Nmax> {List:=List⊕[P];}
```

In particular, if an explicit *oid* (or a `string` value) appears between angle brackets, then its intuitive meaning is that we are interested in selecting the most specific object by means of this *oid* (or this `string` value). On the contrary, if the expression between angle brackets is a query, then this means that we want to select an object, among the instances of the class addressed by the first atom, which satisfy the query. Finally, the instructions in curly brackets store information after each query process.

Descriptors. Let $\mathcal{O}$ be an ontology. A *descriptor* over $\mathcal{O}$, in its most general form, is a rule of the following structure:

$$head :: seq_L \ (seq_M) * \ seq_R$$

---

[5] The symbols in the set $\{\wedge, \vee, \neg\}$ denote the *and*, *or*, and *not* operators, respectively.

where *head* is either an *oid* in Z or a *class atom*, seq_L, seq_M and seq_R are nonempty sequences of atomic body elements. Moreover, valid restrictions on the form of a descriptor can be obtained by omitting $(seq_M)*$ seq_R, or one between seq_L and seq_R. Note that all of the descriptors presented in Section 4.1 comply with this definition. In the following, the left-hand side and right-hand side of a descriptor d are denoted by $head(d)$ and $body(d)$, respectively.

Example 8. We now give a more technical example to show how to recognize strings belonging to the language $csLang = \{a^n b^{2n} c^n : n > 0\}$.[6] We first set the system to use the symbol decomposition described in Section 3. Then, we add the following classes to the ontology:

```
class chain isa object(sym:string, str:string).
class cslan isa object(str:string, n:int).
```

Each instance of `chain` has a string of length one as value for the attribute `sym`, and a nonempty string replicating the value of `sym` as value for the attribute `str`. For example, `ca4:chain(sym:"a", str:"aaaa")` is considered as a valid instance of `chain`. Each instance of `cslan` has a string in $\{a^n b^{2n} c^n : n > 0\}$ as value for the attribute `str` and an integer, representing the number of a's (or of c's) in this string, as value for the attribute `n`. For example, `cs242:cslan(str:"aabbbbcc", n:2)` is considered as valid, while `cs122:cslan(str:"abbcc", n:1)` is not. We now associate one descriptor to each of these classes:

```
<chain(FirstSym, Str)> :: <symbol(value:FirstSym)> {Str:=FirstSym;}
   (<symbol(value:Sym), Sym=FirstSym> {Str:=Str⊕Sym;})*

<cslan(Str, N)> :: <chain("a", Sa)> {Str:=Sa; N:=$length(Sa);}
   <chain("b", Sb), $length(Sb)=2*N> {Str:=Str⊕Sb;}
   <chain("c", Sc), $length(Sc)=N> {Str:=Str⊕Sc;}
```

Regarding the first, note that, since variable `FirstSym` (appearing in the first atomic body element of the first descriptor) is not involved in any recurrence, then its value constrains the rest of the descriptor (see `Sym=FirstSym`). Intuitively, it can be considered as *global*, so that its value can be automatically transferred to the head of the descriptor. On the contrary, variable `Sym`, involved in a recurrence, can be considered as *local*, since its value varies step by step. However, the value of `Sym` can be, at each step, compared to the value of `FirstSym`. For the second descriptor, we just highlight that `$length` is a unary function on strings returning their length.   □

4.3   Semantics

According to our definition of descriptor, we allow four different kinds of bodies, namely seq_L, $seq_L(seq_M)^*$, $seq_L(seq_M)^*seq_R$ and $(seq_M)^*seq_R$. Let $\mathcal{O}$ be an ontology, and d be a descriptor over $\mathcal{O}$. If we consider the body of d as a regular expression over an alphabet of atomic body elements, then we can build from d an

---

[6] The language *csLang* is well known for being a proper context sensitive language.

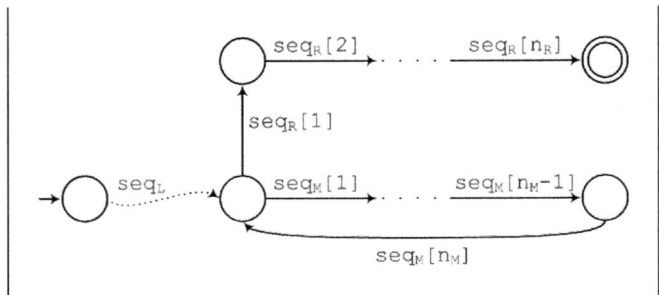

Fig. 2. Finite State Machine for bodies of the form $seq_L(seq_M)^* seq_R$

equivalent *finite state machine*, denoted by $M(d)$. For example, Figure 2 shows the finite state machine associated with a body of the form $seq_L(seq_M)^*$ seq_R.

Note that there are cases where we have inevitably to deal with nondeterminism or ambiguity. In particular, there are two sources of ambiguity. The first, which we term *structural*, may come from the first atomic body elements of both seq_M and seq_R. For example, there is ambiguity if $seq_M[1]$ and $seq_R[1]$ are queries (i) on the same *oid*, (ii) on the same string value, or (iii) on two non-disjoint sets of instances. The second, which we term *semantic*, may arise even if the body of d is of the form seq_L. In fact, during the computation, more than one object (even of the same class) may start from the same position. Section 5 defines syntactic restrictions that limit these form on ambiguity. The following example shows a nontrivial descriptor that is not affected by nondeterminism.

Example 9. Assume to have a string decomposition generating instances of the class natural. An object of this class refers to a text unit of solely digits, the decimal value of which is stored by attribute value. Consider a class dList collecting objects that succinctly represent lists of natural numbers. Each object refers to a list of the form $[n, 2n, 4n, 8n, 16n, \ldots]$, for any $n > 0$. We next define the schemas of these classes and a descriptor, say d_{list}, for class dList.

```
class natural(value:int).

class dList(n:int, size:int).
<dList(N,S)> :: <"["> {S:=0; int Last:=0;}
   <natural(Val)> {N:=Val; S:=S+1; Last:=Val;}
   (<","> <natural(Val), Val=2*Last> {S:=S+1; Last:=Val;})* <"]">
```

Figure 3 shows machine $M(d_{list})$ where the instructions are omitted. Note that $M(d_{list})$ has the same structure of the machine shown in Figure 2, but it is fully deterministic as the strings "," and "]" are different, and since, by definition, class natural cannot have two instances stating from the same position.   □

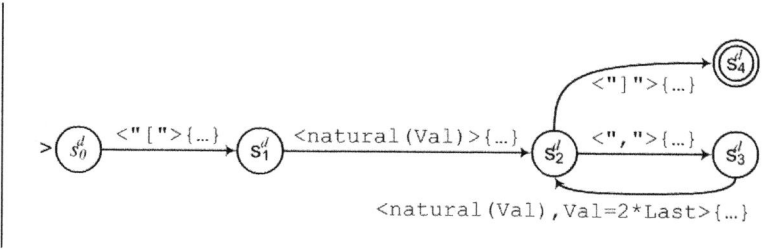

Fig. 3. The deterministic finite state machine $M(d_{list})$

Query Building. We now define how to build queries from a set of descriptors and, in particular, from their associated finite state machines.

First of all, given an ontology $\mathcal{O}$, we need to extend it by some relation schemas and tuples. We call this new ontology $\mathcal{O}'$. More precisely, for each descriptor d such that $x_1, \ldots, x_p$ are the variables appearing in $head(d)$, and $x_{p+1}, \ldots, x_q$ are those appearing only in $body(p)$, with $0 \le p \le q$, we add the following ontology entities to $\mathcal{O}'$:

```
relation config_d(state:int, x1:t1, ..., xq:tq, xa:int, xb:int).
relation initConfig_d(state:int, x1:t1, ..., xq:tq).
          initConfig_d(state:s0d, x1:z1, ..., xq:zq).
```

where attributes $(x_1 : t_1), \ldots, (x_q : t_q)$ refer to the global variables of d, s_0^d is assumed to be the initial state of $M(d)$, and $z_1 \ldots z_q$ are default values assigned to $x_1 \ldots x_q$. Intuitively, in a hypothetical running of $M(d)$, the two schemas allow for maintaining information while $M(d)$ evolves from its initial configuration to some possible final configuration.

Moreover, a set of queries $\mathcal{Q}(d)$ is built from the transitions of $M(d)$. More specifically, for each transition τ of $M(d)$ of the form $(s_i^d, bodyElem, s_j^d)$, where $bodyElem$ is an atomic body element playing the role of transition label, we have to distinguish between two different cases according to whether the left-hand side of $bodyElem$ is either a constant from Z (an *oid* or a string value) or a *query*.

In the first case, let z be the constant value in the left-hand side of $bodyElem$. If s_i^d is either the initial state of $M(d)$ having some ingoing arc, or any state other than the initial one, then we build, from τ, the query

$$config\_d(s_i^d, x_1, \ldots, x_q, x_A, x_B), \ \psi(z, x_B, x_C)$$

where ψ stands for `position` or `str_pos` depending on whether z is an *oid* or a string value, respectively. Moreover, if s_i^d is the initial state of $M(d)$, namely $i = 0$, then we also build the query:

$$initConfig\_d(s_0^d, x_1, \ldots, x_q), \ \psi(z, x_A, x_C)$$

For instance, by considering again the descriptor d_{list} introduced in Example 9, we produce the following three queries from the transitions of $M(d_{list})$ complying with the requirements of this first case:

```
initConfig_d(s0d,N,S,Last), str_pos("[",Xa,Xc)
config_d(s2d,N,S,Last,Xa,Xb), str_pos(",",Xb,Xc)
config_d(s2d,N,S,Last,Xa,Xb), str_pos("]",Xb,Xc)
```

In the second case, let $m_1, \ldots, m_n$ be the query composing the left-hand side of $bodyElem$, where m_1 contains the pair (oid, x_{oid}). We build from τ the query:

$$config\_d(s_i^d, x_1, \ldots, x_q, x_A, x_B), \; position(x_{\text{oid}}, x_B, x_C), \; m_1, \ldots, m_n$$

Moreover, if s_i^d is the initial state of $M(d)$, then we also build

$$initConfig\_d(s_0^d, x_1, \ldots, x_q), \; position(x_{\text{oid}}, x_A, x_C), \; m_1, \ldots, m_n$$

For instance, with respect to Example 9, we produce the following two queries from the transitions of $M(d_{list})$ complying with this second case:

```
config_d(s1d,N,S,Last,Xa,Xb), position(Xoid,Xb,Xc), Xoid:natural(Val)
config_d(s3d,N,S,Last,Xa,Xb), position(Xoid,Xb,Xc), Xoid:natural(Val), Val=2*Last
```

The Yield of a Descriptor. Before computing the yield of a descriptor, we add to $\mathcal{O}'$ all the instances produced by the preprocessing phase and regarding the input decomposition.

Now, given a descriptor d, a *run* of d over $\mathcal{O}'$ consists of evaluating all the queries in $\mathcal{Q}(d)$ over $\mathcal{O}'$ (repeatedly and at least once) in order to extend the ontology itself. We call *run-step* a sequence of evaluations (in a prescribed way) one for each query in $\mathcal{Q}(d)$. Notice that, in general, after evaluating the ith query during a run-step, $\mathcal{O}'$ may not change. However, it may change after evaluating the $(i+1)^{th}$ query. We say that a *run* is *complete* when $\mathcal{O}'$ reaches a fixpoint, that is, when there is no change in $\mathcal{O}'$ after one run-step.

More precisely, suppose that we are evaluating a query created from a transition $(s_i^d, bodyElem, s_j^d)$. As result, for each substitution μ making the query true, we add the following tuple to $\mathcal{O}'$:

```
config_d(sⱼᵈ, z₁, ..., z_q, μ(x_A), μ(x_C))
```

where z_k $(1 \leq k \leq q)$

- is $\mu(x_k)$ if $bodyElem$ contains no instruction assigning the variable x_k (the value of x_k does not change after the current transition); or
- is the result of the unique expression in $bodyElem$ modifying the value of x_k (the value of x_k may change after the current transition).

When a fixpoint has been reached, we look for the tuples of config_d having as first term the unique final state of $M(d)$, say s_{fin}^d. Thus, we extend $\mathcal{O}'$ by adding the following instance:

```
portion(obj:zObj, start:zPos1, end:zPos2).
```

for each tuple of the form

```
config_d(s_{fin}^d, z₁, ..., z_p, ..., z_q, zPos1, zPos2).
```

Algorithm. INSTANCESEEKER($\mathcal{O}, \mathcal{D}, w$)

1. **Input:** $\mathcal{O}$, $\mathcal{D}$, w
2. **Output:** $\mathcal{O}'$
3. $\mathcal{O}' = \mathcal{O}$
4. $Queryes = \emptyset$
5. $FinalStates = \emptyset$
   // Descriptor Compiler:
6. **for each** descriptor $d \in \mathcal{D}$ **do**
   a. $\mathcal{O}' = \mathcal{O}' \cup$ config_Schema(d)
   b. $\mathcal{O}' = \mathcal{O}' \cup$ initConfig_Schema(d)
   c. $\mathcal{O}' = \mathcal{O}' \cup$ initConfig_Tuples(d)
   d. $Queryes = Queryes \cup \mathcal{Q}(d)$
   e. $FinalStates = FinalStates \cup$ finalStatesOf($M(d)$)
   // Document Preprocessor:
7. $\mathcal{O}' = \mathcal{O}' \cup$ documentRepresentationOf(w)
   // Query Executor & Manager:
8. **while** $\mathcal{O}'$ increases in size **do**
   b. **for each** descriptor $d \in \mathcal{D}$ **do**
      ▷ $\mathcal{O}' = \mathcal{O}' \cup$ yield($\mathcal{O}', d$)

Regarding the value $zObj$ we have to distinguish between the descriptor of an object and the descriptor of a class. In the first case, when the head of d is an *oid*, $zObj$ is exactly this *oid*. In the second case, suppose that the head of d is an atom of the class cls with attributes $(\text{oid}, cls), (a_1, t_1), \ldots, (a_p, t_p)$, $p \geq 0$. Each instance of config_d states that we have found a new object of the class cls between positions $zPos_1$ and $zPos_2$, and that the value of its attributes (other than oid) are $z_1, \ldots, z_p$ (recall that $p \leq q$). Therefore we add this new object to $\mathcal{O}'$ (if it is not already present) and $zObj$ is exactly the *oid* of this new object. At this point, a question arises: What is the *oid* of the new instance? We use a function that univocally generates a new *oid* from the class name cls, the values $z_1, \ldots, z_p$ and the positions $zPos_1, zPos_2$. This is in the same spirit of generating different objects for tokens with the same content but appearing in different places. In fact, we want to avoid to generate two *oids* for objects having the same positions, the same attributes and belonging to the same class. We denote by yield($\mathcal{O}', d$) the set of tuples added to $\mathcal{O}'$ after a complete run of d. Finally, given a set $\mathcal{D}$ of descriptors, we compute yield($\mathcal{O}', d$) for each d in $\mathcal{D}$. The algorithm terminates when a *run* reaches a fixpoint.

We now summarize the previous discussion to form a complete algorithm, called INSTANCESEEKER. The algorithm details the entire process that — by starting from a set of descriptors $\mathcal{D}$, an ontology $\mathcal{O}$, and a string w — produces the new ontology $\mathcal{O}'$.

5   Linguistic Restrictions

We point out that the INSTANCESEEKER algorithm may, in general, lead to an infinite loop. For example, the following set of descriptors leads to parse trees of infinite size as well as in *cyclic attribute grammars* [15]:

```
<cls1(X)> :: <token(X)>.
<cls1(X)> :: <cls2(X)>.
<cls2(X)> :: <cls1(Y)> {X:=Y⊕"a"}.
```

Termination is guaranteed if we avoid such cycles. This involves a trivial syntactic check on the set $\mathcal{D}$. However, this restriction is not enough to guarantee usability. Indeed, by only avoiding *cyclicity*, the algorithm requires exponential time in the worst case [15]. In particular, both the number of new objects and the size of attribute values may be exponential. These aspects would seem to make the approach prohibitive. However, such exponential upperbounds are almost never reached in practice. Actually, descriptors employed in real-world applications do not need this power. As an example, the following descriptor shows how attribute values may become of exponential size even if the associated context free grammar has a unique parse tree and only very basic queries are performed:

```
<cls(X,Y,Z)> :: <symbol(Z1)> {X:="0"; Y:="1"; Z:=Z1;}
    (<symbol(Z2)> {X:=X⊕Y; Y:=Y⊕X; Z:=Z⊕Z2;})*.
```

For example, if our algorithm was run on both this descriptor and the string *abcd*, then it would produce the following set of new instances:

```
cls("0","1","a"). cls("0","1","b"). cls("0","1","c"). cls("0","1","d").
cls("01","10","ab"). cls("01","10","bc"). cls("01","10","cd").
cls("0110","1001","abc"). cls("0110","1001","bcd").
cls("01101001","10010110","abcd").
```

Clearly, the size of the first two attributes of `cls` has an exponential growth with respect to the length of the input string. However, in this example, both the role played by the attributes and their composition might be considered somewhat anomalous in the context of IE.

Such empirical considerations have allowed us to identify a number of syntactic restrictions ensuring a proper balance between complexity and expressiveness. Ultimately, we have defined the class $\mathcal{H}$ consisting of pairs of the form $(\mathcal{O}, \mathcal{D})$, where $\mathcal{O}$ is an ontology and $\mathcal{D}$ is a set of descriptors with respect to $\mathcal{O}$. Elements of this class fulfill syntactic restrictions concerning the following aspects:

- the structural interaction between descriptors and ontology,
- the way how attributes can be composed,
- an estimation of the ambiguity that classes, instances and descriptors can have.

5.1   Recursion

Let $\mathcal{O}$ be an ontology, and $\mathcal{D}$ be a set of descriptors with respect to $\mathcal{O}$. Consider the directed dependency graph $G_{[\mathcal{O},\mathcal{D}]} = (N, A)$ built as follows:

1. The set N of nodes contains both the classes and the *oid*s of $\mathcal{O}$, and the data-type `string`.
2. An arc (z, ξ) is in A if z is an *oid* and there is a descriptor $d \in \mathcal{D}$ such that *head*$(d) = z$ and *body*(d) contains an atomic body element starting with ξ.
3. An arc (c, ξ) is in A if c is a class and there is a descriptor $d \in \mathcal{D}$ such that *head*(d) is a class atom starting with c and *body*(d) contains an atomic body element starting with ξ.
4. An arc (c_1, c_2) is in A if c_2 is direct subclass of c_1.
5. An arc (c, ι) is in A if ι is a direct instance of c.

Definition 1. *A pair* $(\mathcal{O}, \mathcal{D})$ *is called* non-recursive *if* $G_{[\mathcal{O},\mathcal{D}]}$ *is cycle free.*   □

5.2   Attribute Composition

Let d be a descriptor. We say that an atomic body element of d, say α, is *strict* if the value of each global variable of d gives a single contribution to the computation of values in the instructions occurring in α. In particular, observe that a variable `X` appearing both in the **then** and in the **else** branches of an arithmetic-if instruction, gives one contribution. The same holds if `X` occurs in such different branches belonging to different arithmetic-if instructions, as long as they have the same if-condition. If a variable occurs only in the condition of arithmetic-if instruction, then its contribution is zero. All other occurrences in the right-hand side of instructions count for one contribution. See [46] for details.

Example 10. Consider the following atomic body elements:

```
<symbol(Z2)> {X:=X⊕Y; Y:=Y⊕X; Z:=Z⊕Z2;}
<symbol(Z2)> {X:=X⊕Y; Y:="a"; Z:=Z⊕Z2;}
<symbol(Z2)> {X:=X⊕"a"; Y:=Y⊕"b"; Z:=Z⊕Z2;}
```

The first one is not strict since both variables `X` and `Y` are used in the right-hand side of two instructions. On the contrary, the second and the third atomic body elements are strict.   □

Definition 2. *Let d be a descriptor, and seq_1 rec seq_2 be its body. We say that d is* strict *if at least one of the following conditions hold:*

- *rec is empty;*
- *rec is of the form* $(seq)^{[h..k]}$ *(where $0 \le h \le k$);*
- *each atomic body element in rec is strict.*   □

5.3 Ambiguity of Classes and Instances

Given a pair $(\mathcal{O}, \mathcal{D})$, we next define the *covering* function ρ associating a set of regexes to each atomic body element, descriptor, instance, and class appearing in $(\mathcal{O}, \mathcal{D})$. Regexes are written according to the JAVA syntax.[7]

Let α be an atomic body element starting with a class of the ontology, say c. The set $\rho(\alpha)$ is defined according to the decomposition technique used for generating instances of c. For example:

- if α starts with `token(value:`*val*`)`, where *val* is a string, then $\rho(\alpha)$ simply contains *val* as regex;
- if α starts with `natural`, then $\rho(\alpha)$ only contains `[0-9]`$^+$;
- if α starts with `token`, then $\rho(\alpha)$ only contains `[\p{Graph}]`$^+$, where `\p{Graph}` stands for any alphanumeric or punctuation character.

Let α be an atomic body element starting with a class of the ontology, say c. Thus, $\rho(\alpha) = \rho(c)$. Let α be an atomic body element starting with an instance of the ontology, say $\hat{\imath}$. Thus, $\rho(\alpha) = \rho(\hat{\imath})$. Let d be a descriptor. The set $\rho(d)$ contains the regex obtained from d by (i) adding the special symbol $\diamond$ among the atomic body elements of d, and (ii) replacing each atomic body element of d, say α, with $\rho(\alpha)$. For example, if d is the descriptor of class `moderateVar` defined in Section 4.1, then the set $\rho(d)$ contains the following regex:

$$\texttt{High:}\diamond\texttt{[\textbackslash p\{Graph\}]}^+(\diamond\texttt{[\textbackslash p\{Graph\}]}^+)^*\diamond\texttt{Low:}\diamond\texttt{[\textbackslash p\{Graph\}]}^+$$

Let $\hat{\imath}$ be an instance, and $d_1, \ldots, d_h$ be only and all the descriptors of $\hat{\imath}$. The set $\rho(\hat{\imath})$ consists of the regexes $\rho(d_1), \ldots, \rho(d_h)$. Let c be a class. The set $\rho(c)$ consists of the regexes $\rho(d_1), \ldots, \rho(d_h), \rho(\hat{\imath}_1), \ldots, \rho(\hat{\imath}_k), \rho(c_1), \ldots, \rho(c_q)$, where $d_1, \ldots, d_h, \hat{\imath}_1, \ldots, \hat{\imath}_k$, and $c_1, \ldots, c_q$ are only and all the descriptors, the direct instances of, and the direct subclasses of c, respectively. Two regexes are called *concurrent* if the intersection of the languages they define is nonempty.

Definition 3. *A class c is called* ambiguous *if at least one of the following conditions hold:*

- *$\rho(c)$ contains two concurrent regexes;*
- *c has an ambiguous direct subclass;*
- *c has a descriptor with an atomic body element starting with an ambiguous class.*   □

Definition 4. *A class or an instance ξ is called* strongly ambiguous *if at least one of the following conditions hold:*

- *ξ has a descriptor with a recurrence part containing an atomic body element starting with an ambiguous class;*
- *ξ has a descriptor with an atomic body element which starts with a strongly ambiguous class or instance;*

---

[7] See `http://download.oracle.com/javase/6/docs/api/java/util/regex/`
`Pattern.html`

– ξ *is a class either being a superclass of a strongly ambiguous class or having*
 a strongly ambiguous instance.   □

As a remark, no class nor instance previously defined is strongly ambiguous.
Moreover, only class `wDay` is simply ambiguous since their instances contain concurrent regexes, namely `Today` and `Tomorrow`. Finally, in the following example class `c` is ambiguous, and instance `inst` is strongly ambiguous:

```
<c(X)> :: <token(X)>.
<c(X)> :: <natural(N)> {X:=$toStr(N);}.
<inst> :: <token> (<c>)+.
```

5.4   Ambiguity of Descriptors

In this section we consider the problem of checking whether a descriptor has to be considered ambiguous or not. For instance, consider the following descriptor:

```
<c(C1,C2)> :: <C1:cls1> <C2:cls2>.
```

Now, assume that $\rho(\text{cls1}) = \text{a(bc)}^*$, $\rho(\text{cls2}) = \text{(bc)}^*\text{a}$, and the input string is abca. Clearly, this string can be decomposed in a bca but also in abc a. In this particular case, two objects of a class are recognized in one portion of text, and in general this may lead in recognizing several objects of a class in one portion of text.

To preserve the property that each descriptor generates at most one new object per portion, we introduce the following restriction. To this aim, we use the *covering* function ρ for establishing whether a descriptor may lead to ambiguity. Let r_1 and r_2 be two regular expressions. Consider the following algorithm that builds an NFA $A(r_1, r_2)$ associated with the pair (r_1, r_2):

1. Let A_1 be the minimal DFA of r_1;
2. add a new state, say s_1, to A_1;
3. set s_1 as the unique initial state of A_1;
4. add an ε-transitions from s_1 to each final state of A_1;
5. modify A_1 in such a way that it does not recognize the empty string;
6. let A_2 be an NFA of r_2;
7. let $A(r_1, r_2) = (A_1 \circ A_2) \cap A_2$.

Definition 5. *A pair of regular expressions (r_1, r_2) is ambiguous if the language defined by NFA $A(r_1, r_2)$ is nonempty.*   □

Definition 6. *A descriptor is ambiguous if it contains two consecutive chains of atomic body elements, say γ_1 and γ_2, such that the pair $(\rho(\gamma_1), \rho(\gamma_2))$ is ambiguous.*   □

5.5   The Class $\mathcal{H}$

We now define the class $\mathcal{H}$ consisting of pairs of the form $(\mathcal{O}, \mathcal{D})$, where $\mathcal{O}$ is an ontology and $\mathcal{D}$ is a set of descriptors with respect to $\mathcal{O}$. In particular, a pair $(\mathcal{O}, \mathcal{D})$ belongs to $\mathcal{H}$ if all the following restrictions are satisfied:

- the pair $(\mathcal{O}, \mathcal{D})$ is *non-recursive*;
- each *non-strict* descriptor in $\mathcal{D}$ only uses operators form $\{|\ |, +, -, \div, \%\}$ in its recurrence part;
- classes and instances of $\mathcal{O}$ are not *strongly ambiguous* with respect to $\mathcal{D}$;
- each descriptor in $\mathcal{D}$ is not *ambiguous*.

5.6   Complexity Issues

Let $\mathcal{O}$ be an ontology, $\mathcal{D}$ be a set of descriptors with respect to $\mathcal{O}$, and w be a string. We next study the data complexity of the INSTANCESEEKER algorithm subject to the following assumptions:

1. the pair $(\mathcal{O}, \mathcal{D})$ belongs to $\mathcal{H}$;
2. both $\mathcal{D}$ and the schemas of $\mathcal{O}$ are fixed.

Lemma 1. *If the pair $(\mathcal{O}, \mathcal{D})$ belongs to class $\mathcal{H}$, then the size of each new object generated by the INSTANCESEEKER algorithm is polynomial in the size of w.*

Proof (sketch). This directly follows by combining the results obtained in [46] with the fact that both the pair $(\mathcal{O}, \mathcal{D})$ is non-recursive and each *non-strict* descriptor in $\mathcal{D}$ only uses operators form $\{|\ |, +, -, \div, \%\}$ in its recurrence part.

Lemma 2. *Let $\mathcal{O}$ be an ontology, and $\mathcal{Q}$ be a query on $\mathcal{O}$. If the schemas of $\mathcal{O}$ are fixed, then $ans_{\mathcal{Q}}(\mathcal{O})$ can be computed in time polynomial in the size of $\mathcal{O}$.*

Proof (sketch). Since arities are fixed, the number of substitutions to be considered is polynomial, and checking whether a substitution belongs to $ans_{\mathcal{Q}}(\mathcal{O})$ only requires liner time.

Lemma 3. *If the pair $(\mathcal{O}, \mathcal{D})$ belongs to class $\mathcal{H}$, then the INSTANCESEEKER algorithm generates, for each portion, at most one instance per class.*

Proof (sketch). Since classes and instances of $\mathcal{O}$ are not strongly ambiguous with respect to $\mathcal{D}$, and each descriptor is not ambiguous, then no portion can be associated with two existing instances of the same class (other than `object`).

Theorem 1. *If the pair $(\mathcal{O}, \mathcal{D})$ belongs to class $\mathcal{H}$, then the INSTANCESEEKER algorithm runs in polynomial time with respect to the size of the pair $(\mathcal{O}, w)$.*

Proof (sketch). The result follows by Lemmas 1, 2 and 3.

6   System and Applications

The HɪLɛX system, as a result of a positive feedback from business customers, has been industrialized for its commercial distribution by Exeura Srl, a technology company working on analytics, data mining, and knowledge management.

In this section, after a brief panorama of the system architecture and regarding some implementation principles, we describe some real-world applications developed in collaboration with Exeura Srl, namely the *iTravel System*, and other information extraction applications from *balance sheets*, *news*, and *clinical documents*.

6.1   System Architecture

A simplified architecture of the HɪLɛX system, implementing the semantic information extraction approach described in the previous sections, is represented in Figure 4.

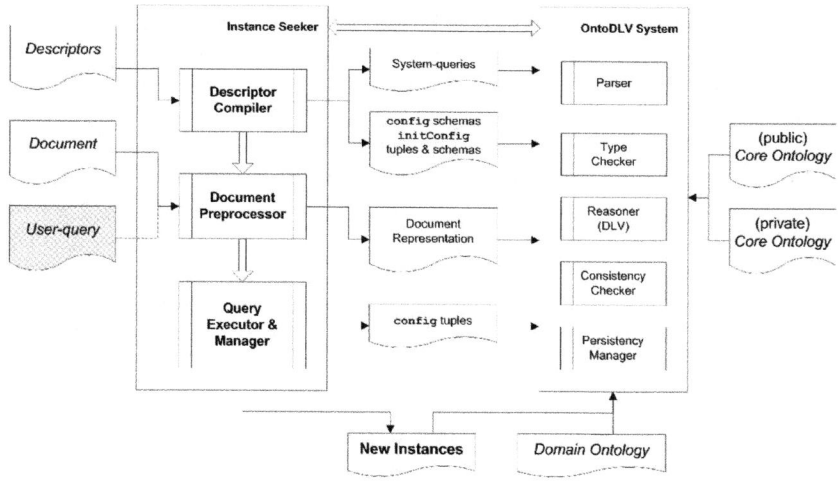

Fig. 4. HɪLɛX Core Diagram

The information extraction process is executed in three main steps: compiling, preprocessing and running. The input of the whole process is composed by an unstructured document, an ontology and a set of descriptors. Optionally, also a user query can be initially specified. The output consists of instances representing objects found or discovered in the document.

The first step, performed by the *Descriptor Compiler* module, translates the semantic rules into OntoDLP queries. Moreover, for each descriptor d, the module also produces the schemas config_d and initConfig_d, and the tuples for the second relation. Actually, each descriptor is not translated in a set of queries as described in the previous section, but it is translated to an ASP logic program whose bodies are the aforementioned defined queries and whose heads are `config_d` or `position` atoms.

The second step, performed by the *Document Preprocessor* module, processes the input document to obtain its ontological representation. In particular, its output contains instances for the schemas `position` and `str_pos`, and for the schemas `token`, `word`, `symbol`, `email`, `date`, and possibly others. This module integrates different external tools for the document decomposition which are enabled or disabled depending on the objects exploited by the descriptors.

Finally, the third step is performed by the *Query Executor & Manager* module and orchestrates the execution of the reasoning engine *DLV*, a state-of-the-art ASP system [43].

6.2 An E-Tourism Application

iTravel [34, 54] is an e-tourism system developed under PIA (Pacchetti Integrati di Agevolazione industria, artigianato e servizi) project "iTravel: Intelligent Tourism Advisor' funded by the Calabrian Region.[8] The iTravel system helps both employees and customers of a travel agency in finding the best possible travel solution in a short time. It can be seen as a "mediator" system finding the best match between the offers of the tour operators and the requests of the tourists.

iTravel, like other existing portals, has been equipped with a proper (web-based) user interface; but, behind the user interface, there is an "intelligent" core that exploits knowledge representation and reasoning technologies based on ASP. In iTravel, the information regarding the touristic offers provided by tour operators is received by the system as a set of emails. Each e-mail might contain plain text and/or a set of leaflets, usually distributed as pdf or image files, which store the details of the offer (e.g., place, accommodation, price etc.). Leaflets are designed to be human-readable, might mix text and images, and usually do not have a uniform layout. Emails (and their content) are automatically processed by using the H*i*L*ε*X system, and the extracted data about touristic offers is used to populate an OntoDLP ontology that models the domain of discourse, namely the *"Tourism Ontology"*. The resulting ontology is then analyzed by using a set of reasoning modules combining the extracted data with the knowledge regarding places (geographical information) and users (user preferences). The system mimics the typical deductions made by a travel agency employee for selecting the most appropriate answers to the user needs. We now briefly describe the Tourism Ontology and the implementation of the ASP-based features of iTravel.

The Tourism Ontology has been specified by analyzing the nature of the input in cooperation with the staff of a real touristic agency. This way, the key

---

[8] The project team involved four organizations: the Department of Mathematics of the University of Calabria (which has ASP as one of the principal research area), Exeura Srl (a company working on knowledge management), Top Class Srl (a travel agency), and ASPIdea (a software farm specialized in the development of web applications).

```
class place (description:string).
class transportMean (description:string).
class tripKind (description:string).
class customer (firstName:string, lastName:string, birthDate:date,
    status:string, childNumber:int, job:job).
class touristicOffer (start:place, destination:place, kind:tripKind,
    means:transportMean, cost:int, fromDay:date,
    toDay:date, maxDuration:int, deadline:date, uri:string).
relation placeOffer (place:place, kind:tripKind).
```

Fig. 5. Some entities from the Tourism Ontology

entities that describe the process of organizing and selling a complete holiday package could be modeled. In particular, the Tourism Ontology models all the required information, such as geographic information, kind of holiday, transportation means, etc. In Figure 5, we report some entities constituting the Ontology. In detail, class Customer allows us to model the personal information of each customer. The kind of trip is represented by using a class tripKind. Examples of tripKind instances are: safari, sea_holiday, etc. In the same way, e.g., airplane, train, etc., are instances of class transportMean. Geographical information is modeled by means of class Place, which has been populated by exploiting Geonames[9], one of the largest publicly-available geographical databases. Moreover, each place is associated with a kind of trip by means of the relation placeOffer (e.g., Kenya offers safari, Sicily offers both sea and sightseeing).

The mere geographic information is then enriched by other information that is usually exploited by travel agency employees for selecting a travel destination. For instance, one might suggest avoiding sea holidays in winter; whereas, one should be recommended a visit to Sicily in summer. Finally, the touristicOffer class contains an instance for each available holiday package. The instances of this class are added either manually by the personnel of the agency, are automatically by the H*ι*L*ε*X system. In the last case, we have used descriptors such as the following:

```
<touristicOffer(destination:D,fromDay:FD,toDay:TD)> ::
    <D:place("Sicily")>  <token>*  <FD:date>  <token>*  <TD:date>.
```

Finally, the *personalized trip search* feature has been conceived to simplify the task of selecting the holiday packages that best fit the customer needs. In a typical scenario, when a customer enters the travel agency, what has to be clearly understood (for properly selecting a holiday package fitting the customer needs) is summarized in the following four words: *where, when, how,* and *budget.* However, the customer does not directly specify all this information, for example, he can ask for a sea holiday in January but he does not specify a precise place, or he can ask for a kind of trip that is unfeasible in a given period. In iTravel, current needs are specified by filling an appropriate search form, where some of the key information has to be provided (i.e., where and/or when and/or available

---

[9] See www.geonames.org

money and/or how). Note that the Tourism Ontology models the knowledge of the travel agent. Moreover, the extraction process continuously populates the ontology with new touristic offers. Thus, the system, by running a specifically devised reasoning module, combines the specified information with the one available in the ontology, and shows the holiday packages that best fit the customer needs.

6.3 Other Applications

IE from Balance Sheets. A *balance sheet* is a snapshot of a business' financial condition at a specific moment in time, usually at the close of an accounting period. A balance sheet comprises assets, liabilities, and owners' or stockholders' equity. Balance sheets contain an unstructured "additional notes" section. The additional notes section contains expenses categories and their costs (research and development, government taxes, etc.). The problem is that each company has a different way to represents the expenses.

An interesting task (see Figure 6) is to automatically extract expense categories and their values from the additional notes section. The recognition process must be independent from the position of the data in the document. Moreover, different categories should be identified on the basis of their concept (i.e. "R&D" and "Research & Dev." are equivalent). Finally, extracted information has to be loaded into a relational database to perform additional statistical analysis or data synthesis tasks.

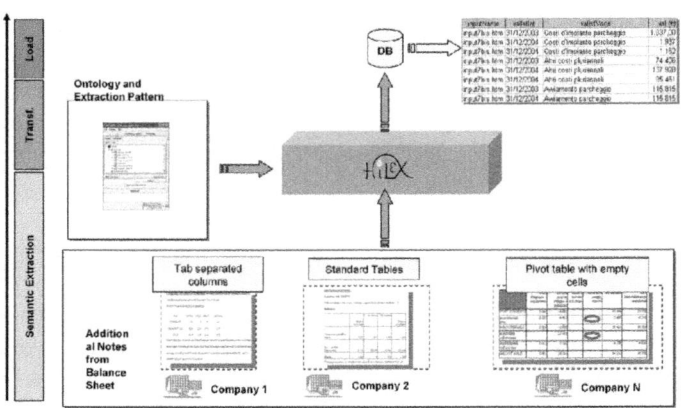

Fig. 6. HιLεX at work on Balance Sheets

IE from News. *CODECISION* is an ICT company which produces decision support solutions based on information intelligence techniques. In this context CODECISION was interested in analyzing financial market news. Financial market news cover concepts related to sales, stock indexes, market agreements, restructuring and financial problems of companies and so on. The analysis of these news represents, for many companies, an interesting instrument for decision support purposes. For example, for a company in crisis could be useful to extract from news information about stock increases, sales increase, competitors in bad situation etc. In this case the desired results is to provide a "semantic" extraction of information from news, and to recognize and to extract the relevant concepts even if they are expressed in different forms.

IE from Clinical Documents. Exchanging clinical information is known to be a hard problem. Patient information is generally distributed through out different, heterogeneous, structured and unstructured clinical documents (letters, reports, forms). However, data exchange standards for clinical documents have been defined by the health authorities to reach a fully structured coded communication.

In this scenario a desired result would be to assist healthcare professionals in executing and monitoring clinical processes by providing functionalities for automatic knowledge acquisition. Moreover, it is important to produce structural clinical documents from unstructured information contained in existing information systems.

The H*i*L*ε*X system enables the recognition of semantic patterns representing concepts of the medical domain (see [56] for further detail). Thus, it is possible to acquire automatically medical knowledge from unstructured clinical documents. This way, patient healthcare information is contained in documents according to defined standardsx, and acquired information can be analyzed for identifying main causes of medical errors, high costs and, potentially, to suggest clinical processes restructuring or improvement able to enhance cost control and patient safety.

7   Conclusion and Future Work

This work presents a novel, concrete, powerful and expressive approach to information extraction from non-structured documents. The approach, implemented in the H*i*L*ε*X system, is founded on the following ideas:

- a novel extraction language for specifying semantic rules to discover and extract objects from non-structured documents;
- a unified document representation;
- the use of ontologies for representing the documents and the domain of interest;
- modularity and extensibility with respect to existing techniques to IE.

As far as the main contributions of this work are concerned, we have:

- the formal definition of both syntax and semantics of the extraction language;
- an overview of the complexity analysis;
- design and implementation of a system founded on the defined language;
- a survey of some real-world applications.

Finally, both the language and the system are still under developmental. Current and new issues to be considered include:

- optimized techniques for improving the overall performance on very large sets of documents;
- extensions of the structure of descriptors with no loss of efficiency;
- software "wizards" for supporting the user during the definition and the management of both ontologies and descriptors.

References

1. Adelberg, B.: NoDoSE – a tool for semi-automatically extracting structured and semistructured data from text documents. SIGMOD Rec. 27(2), 283–294 (1998)
2. Agichtein, E., Gravano, L.: Snowball: extracting relations from large plain-text collections. In: Proceedings of DL 2000, San Antonio, Texas, United States, pp. 85–94. ACM, New York (2000)
3. Arocena, G.O., Mendelzon, A.O.: WebOQL: restructuring documents, databases, and webs. Theor. Pract. Object Syst. 5(3), 127–141 (1999)
4. Banko, M., Cafarella, M.J., Soderland, S., Broadhead, M., Etzioni, O.: Open information extraction from the web. In: Proceedings of IJCAI 2007, Hyderabad, India, pp. 2670–2676. Morgan Kaufmann Publishers Inc., San Francisco (2007)
5. Brin, S.: Extracting Patterns and Relations from the World Wide Web. In: Atzeni, P., Mendelzon, A.O., Mecca, G. (eds.) WebDB 1998. LNCS, vol. 1590, pp. 172–183. Springer, Heidelberg (1999)
6. Califf, M.E., Mooney, R.J.: Relational learning of pattern-match rules for information extraction. In: Proceedings of AAAI 1999/IAAI 1999, Orlando, Florida, United States, pp. 328–334. American Association for Artificial Intelligence, Menlo Park (1999)
7. Carme, J., Gilleron, R., Lemay, A., Niehren, J.: Interactive learning of node selecting tree transducer. Machine Learning 66(1), 33–67 (2007)
8. Chang, C.-H., Hsu, C.-N., Lui, S.-C.: Automatic information extraction from semi-structured Web pages by pattern discovery. Decis. Support Syst. 35(1), 129–147 (2003)
9. Chang, C.-H., Kayed, M., Girgis, M.R., Shaalan, K.F.: A Survey of Web Information Extraction Systems. IEEE Trans. on Knowl. and Data Eng. 18(10), 1411–1428 (2006)
10. Cimiano, P., Handschuh, S., Staab, S.: Towards the self-annotating web. In: Proceedings of WWW 2004, pp. 462–471. ACM, New York (2004)
11. Crescenzi, V., Mecca, G.: Grammars have exceptions. Inf. Syst. 23(9), 539–565 (1998)
12. Crescenzi, V., Mecca, G.: Automatic information extraction from large websites. J. ACM 51(5), 731–779 (2004)
13. Crescenzi, V., Mecca, G., Merialdo, P.: Roadrunner: Towards automatic data extraction from large web sites. In: Proceedings of VLDB 2001, pp. 109–118. Morgan Kaufmann Publishers Inc., San Francisco (2001)
14. de Bruijn, J., Martin-Recuerda, F., Manov, D., Ehrig, M.: State-of-the-art survey on Ontology Merging and Aligning v1. Technical report, SEKT project deliverable D4.2.1 (2004),
http://sw.deri.org/~jos/sekt-d4.2.1-mediation-survey-final.pdf

15. Efremidis, S., Papadimitriou, C.H., Sideris, M.: Complexity characterizations of attribute grammar languages. Inf. Comput. 78(3), 178–186 (1988)
16. Eikvil, L.: Information Extraction from World Wide Web - A Survey. Technical Report 945, Norweigan Computing Center (1999)
17. Embley, D.W.: Towards Semantic Understanding – An Approach Based on Information Extraction Ontologies. In: Proceedings of ADC 2004, Dunedin, New Zealand. Database Technologies, vol. 27 (2004)
18. Embley, D.W., Campbell, D.M., Jiang, Y.S., Liddle, S.W., Lonsdale, D.W., Ng, Y.-K., Smith, R.D.: Conceptual-model-based data extraction from multiple-record web pages. Data Knowl. Eng. 31(3), 227–251 (1999)
19. Embley, D.W., Jiang, Y.S., Ng, Y.-K.: Record-boundary discovery in web documents. In: SIGMOD Conference, pp. 467–478 (1999)
20. Embley, D.W., Lopresti, D., Nagy, G.: Notes on Contemporary Table Recognition. In: Bunke, H., Spitz, A.L. (eds.) DAS 2006. LNCS, vol. 3872, pp. 164–175. Springer, Heidelberg (2006)
21. Etzioni, O., Cafarella, M., Downey, D., Kok, S., Popescu, A.-M., Shaked, T., Soderland, S., Weld, D.S., Yates, A.: Web-scale information extraction in knowitall (preliminary results). In: Proceedings of WWW 2004, pp. 100–110. ACM, New York (2004)
22. Feldman, R., Aumann, Y., Finkelstein-Landau, M., Hurvitz, E., Regev, Y., Yaroshevich, A.: A Comparative Study of Information Extraction Strategies. In: Gelbukh, A. (ed.) CICLing 2002. LNCS, vol. 2276, pp. 349–359. Springer, Heidelberg (2002)
23. Feldman, R., Rosenfeld, B., Fresko, M.: TEG – a hybrid approach to information extraction. Knowledge and Information Systems 9(1), 1–18 (2006)
24. Freitag, D.: Information extraction from HTML: application of a general machine learning approach. In: Proceedings of AAAI 1998/IAAI 1998, Madison, Wisconsin, United States, pp. 517–523. American Association for Artificial Intelligence, Menlo Park (1998)
25. Freitag, D.: Machine learning for information extraction in informal domains. Machine Learning 39(2), 169–202 (2000)
26. Gruber, T.R.: A translation approach to portable ontology specifications. Knowl. Acquis. 5(2), 199–220 (1993)
27. Gruber, T.R.: Toward principles for the design of ontologies used for knowledge sharing. Int. J. Hum.-Comput. Stud. 43(5-6), 907–928 (1995)
28. Guarino, N.: Formal ontology and information systems. In: International Conference On Formal Ontology In Information Systems FOIS 1998, Trento, ITALY, pp. 3–15. IOS Press, Amsterdam (1998)
29. Hammer, J., García-Molina, H., Nestorov, S., Yerneni, R., Breunig, M., Vassalos, V.: Template-based wrappers in the TSIMMIS system. SIGMOD Rec. 26(2), 532–535 (1997)
30. Hammer, J., McHugh, J., Garcia-Molina, H.: Semistructured Data: The Tsimmis Experience. In: Proceedings of ADBIS 1997, St.-Petersburg, Nevsky Dialect, pp. 1–8 (1997)
31. Hassan, T., Baumgartner, R.: Table recognition and understanding from pdf files. In: Proceedings of ICDAR 2007, pp. 1143–1147. IEEE Computer Society, Washington, DC (2007)
32. Hsu, C.-N., Dung, M.-T.: Generating finite-state transducers for semi-structured data extraction from the Web. Inf. Syst. 23(9), 521–538 (1998)
33. Hu, J., Kashi, R., Lopresti, D., Wilfong, G.: Experiments in table recognition (2001)

34. Ielpa, S.M., Iiritano, S., Leone, N., Ricca, F.: An ASP-Based System for e-Tourism. In: Erdem, E., Lin, F., Schaub, T. (eds.) LPNMR 2009. LNCS, vol. 5753, pp. 368–381. Springer, Heidelberg (2009)
35. Kieninger, T., Dengel, A.R.: The T-Recs Table Recognition and Analysis System. In: Lee, S.-W., Nakano, Y. (eds.) DAS 1998. LNCS, vol. 1655, pp. 255–270. Springer, Heidelberg (1999)
36. Knuth, D.E.: Semantics of context-free languages. Theory of Computing Systems 2(2), 127–145 (1968)
37. Kuhlins, S., Tredwell, R.: Toolkits for Generating Wrappers. In: Aksit, M., Awasthi, P., Unland, R. (eds.) NODe 2002. LNCS, vol. 2591, pp. 184–198. Springer, Heidelberg (2003)
38. Kushmerick, N.: Wrapper induction: efficiency and expressiveness. Artif. Intell. 118(1-2), 15–68 (2000)
39. Kushmerick, N., Thomas, B.: Adaptive Information Extraction: Core Technologies for Information Agents. In: Klusch, M., Bergamaschi, S., Edwards, P., Petta, P. (eds.) Intelligent Information Agents. LNCS (LNAI), vol. 2586, pp. 79–103. Springer, Heidelberg (2003)
40. Kushmerick, N., Weld, D.S., Doorenbos, R.B.: Wrapper Induction for Information Extraction. In: Proceedings of IJCAI 1997, NAGOYA, Aichi, Japan, pp. 729–737 (1997)
41. Laender, A.H.F., Ribeiro-Neto, B., da Silva, A.S.: DEByE - Data Extraction By Example. Data Knowl. Eng. 40(2), 121–154 (2002)
42. Laender, A.H.F., Ribeiro-Neto, B.A., da Silva, A.S., Teixeira, J.S.: A brief survey of web data extraction tools. SIGMOD Rec. 31(2), 84–93 (2002)
43. Leone, N., Pfeifer, G., Faber, W., Eiter, T., Gottlob, G., Perri, S., Scarcello, F.: The dlv system for knowledge representation and reasoning. ACM Trans. Comput. Logic 7(3), 499–562 (2006)
44. Liu, B., Grossman, R., Zhai, Y.: Mining Web Pages for Data Records. IEEE Intelligent Systems 19(6), 49–55 (2004)
45. Liu, L., Pu, C., Han, W.: XWRAP: An XML-Enabled Wrapper Construction System for Web Information Sources. In: Proceedings of ICDE 2000, San Diego, CA, USA, pp. 611–621. IEEE Computer Society, Washington, DC (2000)
46. Manna, M., Scarcello, F., Nicola, L.: On the complexity of regular-grammars with integer attributes. J. Comput. System Sci., 1–29 (2010)
47. Mecca, G., Atzeni, P., Masci, A., Sindoni, G., Merialdo, P.: The Araneus Web-based management system. SIGMOD Rec. 27(2), 544–546 (1998)
48. Muslea, I., Minton, S., Knoblock, C.: A hierarchical approach to wrapper induction. In: Proceedings of AGENTS 1999, Seattle, Washington, United States, pp. 190–197. ACM, New York (1999)
49. Muslea, I., Minton, S., Knoblock, C.A.: Hierarchical Wrapper Induction for Semistructured Information Sources. Autonomous Agents and Multi-Agent Systems 4(1-2), 93–114 (2001)
50. Pivk, A., Cimiano, P., Sure, Y., Gams, M., Rajkovič, V., Studer, R.: Transforming arbitrary tables into logical form with TARTAR. Data Knowl. Eng. 60(3), 567–595 (2007)
51. Pivk, E., Sure, Y.: From tables to frames. Journal of Web Semantics, 166–181 (2005)
52. Predoiu, L., de Bruijn, J., Feier, C., Scharffe, F., Martín-Recuerda, F., Manov, D., Ehrig, M.: State-of-the-art survey on ontology merging and aligning v2. Deliverable D4.2.2, SEKT (2005)

53. Ribeiro-Neto, B., Laender, A.H.F., da Silva, A.S.: Extracting semi-structured data through examples. In: Proceedings of CIKM 1999, Kansas City, Missouri, United States, pp. 94–101. ACM (1999)
54. Ricca, F., Alviano, M., Dimasi, A., Grasso, G., Ielpa, S.M., Iiritano, S., Manna, M., Leone, N.: A Logic-Based System for e-Tourism. Fundamenta Informaticae 105, 35–55 (2010)
55. Ricca, F., Leone, N.: Disjunctive logic programming with types and objects: The dlv$^+$ system. J. Applied Logic 5(3), 545–573 (2007)
56. Ruffolo, M., Manna, M., Cozza, V., Ursino, R.: Semantic clinical process management. In: CBMS, pp. 518–523 (2007)
57. Sahuguet, A., Azavant, F.: Building intelligent Web applications using lightweight wrappers. Data Knowl. Eng. 36(3), 283–316 (2001)
58. Soderland, S.: Learning Information Extraction Rules for Semi-Structured and Free Text. Mach. Learn. 34(1-3), 233–272 (1999)
59. Wu, F., Weld, D.S.: Autonomously semantifying wikipedia. In: Proceedings of CIKM 2007, Lisbon, Portugal, pp. 41–50. ACM, New York (2007)
60. Yildiz, B., Kaiser, K., Miksch, S.: pdf2table: A method to extract table information from pdf files. In: IICAI, pp. 1773–1785 (2005)
61. Zanibbi, R., Blostein, D., Cordy, J.R.: A survey of table recognition: Models, observations, transformations, and inferences. Int'l J. Document Analysis and Recognition 7, 1–16 (2004)

DSToolkit: An Architecture for Flexible Dataspace Management[*]

Cornelia Hedeler[1], Khalid Belhajjame[1], Lu Mao[1], Chenjuan Guo[1],
Ian Arundale[1], Bernadette Farias Lóscio[2], Norman W. Paton[1],
Alvaro A.A. Fernandes[1], and Suzanne M. Embury[1]

[1] School of Computer Science, The University of Manchester
Oxford Road, Manchester M13 9PL, UK
{chedeler,khalidb,maol,guoc,arundai7,norm,alvaro,
embury}@cs.manchester.ac.uk
[2] Universidade Federal de Pernambuco, Centro de Informatica
Cidade Universitria 50740-540, Recife, PE, Brasil
bfl@cin.ufpe.br

Abstract. The vision of dataspaces is to provide various of the benefits
of classical data integration, but with reduced up-front costs. Combining
this with opportunities for incremental refinement enables a 'pay-as-you-
go' approach to data integration, resulting in simplified integrated access
to distributed data. It has been speculated that model management could
provide the basis for Dataspace Management, however, this has not been
investigated until now.

Here, we present DSToolkit, the first dataspace management system
that is based on model management, and therefore, benefits from the
flexibility provided by the approach for the management of schemas
represented in heterogeneous models, supports the complete dataspace
lifecycle, which includes automatic initialisation, maintenance and improvement of a dataspace, and allows the user to provide feedback by
annotating result tuples returned as a result of queries the user has posed.
The user feedback gathered is utilised for improvement by annotating, selecting and refining mappings. Without the need for additional feedback
on a new data source, these techniques can also be applied to determine
its perceived quality with respect to already gathered feedback and to
identify the best mappings over all sources including the new one.

Keywords: Dataspace Management System, Dataspace lifecycle,
Incremental improvement.

1 Introduction

1.1 Motivation

Data integration in various forms has been the focus of ongoing research in the
database community for over 20 years. The objective is to provide an integrated

---

[*] The work reported in this paper was supported by a grant from the EPSRC.

A. Hameurlain et al. (Eds.): TLDKS V, LNCS 7100, pp. 126–157, 2012.

view and access to multiple heterogeneous data sources. Typically this involves developing a single global integration schema to which the schemas of the sources are related by some form of mapping. Using those mappings, queries posed over the central schema are then unfolded [24], optimised by a distributed query processor and evaluated over the sources. Various systems have been developed that are based on the approach of classical data integration, e.g., DB2 [23] or OpenII [45]. However, defining and maintaining mappings between a global integration schema and various source schemas has proven to be labour intensive [13], in particular in a world of ever more data sources that evolve over time to account for the changes in the data to be stored. This means that classical data integration is most effective when integrating small numbers of stable data sources, but less so for the integration of large numbers of evolving data sources, or for on-the-fly data integration.

The vision of *dataspaces* [19,25] aims to realise various of the benefits of classical data integration but with much lower startup costs thereby supporting integration on demand but with lower quality of the resulting integration. The quality of the integration can then be improved over time in a 'pay-as-you-go' manner utilising feedback provided by the user or developer. Various dataspace management systems have been proposed in recent years, e.g., SEMEX [18], iMeMex [15], PayGo [35], and Q [48].

Bernstein *et al.* [8] speculated that model management could form the basis of dataspace management, but so far this has not been investigated as none of the dataspace management systems proposed so far are based on model management, and nor have any of the existing model management systems (e.g., MISM [1], Rondo [39], GeRoMe [31] or Automed [46]) been extended to support the incremental improvement that is integral to dataspace management systems.

Dataspace systems can vary in numerous dimensions [27,26], but as illustrated in the surveys [27,26] the dataspace systems proposed so far tend to be somewhat narrowly focussed, targeting specific applications and making assumptions that do not hold elsewhere, e.g., SEMEX [18], a personal information management system, requires domain knowledge to identify associations between schemas which may not always be available, iMeMex, also managing personal information, requires path-based queries called *trails* [15] that are provided manually and require the user to have a good understanding of the schemas and the relationships between them, a requirement that shuts out the casual user, and PayGo [35], integrating web sources, only supports a union schema as an integration schema.

In this paper we present DSToolkit, an architecture for flexible dataspace management that is based on model management, thereby benefiting from the flexibility of being able to manage multiple heterogeneous models, that supports the whole lifecycle of a dataspace [26] including automatic support for initialisation, also called bootstrapping, but also provides support for maintenance in the context of source changes and that provides a means for incremental improvement of the dataspace by the casual user.

1.2   Overview of the Approach

DSToolkit builds on the foundations provided by model management systems, in particular MISM [1] in the sense that we use a model-independent super-model to capture schemas represented by heterogeneous models. Furthermore, *DSToolkit* contains implementations of various model management operators [7,6,8] to support the management of multiple schemas. Model management operators implemented include `match`, `merge`, `difference` and `compose`. Utilising the results of model management research enables us to provide a system that is flexible in its management of multiple heterogeneous schemas and the associations between constructs in those schemas and to support the maintenance of the dataspace. For example, constructs can be attributes or tables in relational schemas, or simple or complex elements in XML schemas.

However, various changes and extensions are necessary to turn a model management system into a dataspace management system:

1. We have generalised the supermodel even further to emphasise the common-alities in terms of the role the constructs play in the various models (e.g., whether they represent relationships between other constructs, can have val-ues, or contain other constructs), rather than their differences. The more specific information on their differences is required for ModelGen [1], but only to a certain extent needed to be able to express relationships between constructs or to query the underlying data sources.

2. We have introduced various kinds of *morphisms*, which are not as integral to MISM as they are to other model management platforms, such as Rondo [39]. The morphisms we introduce represent associations between elements in dif-ferent schemas at different levels of abstraction, and are *matches*, *schematic correspondences* and *mappings*. *Matches* give an indication that two con-structs are similar to a certain extend and tend to be the relationships re-turned by matching algorithms. *Schematic correspondences* [36] are based on schematic heterogeneities introduced in [33,32] and represent richer semantic relationships between schema elements, such as same or different names for the same construct, missing constructs (e.g., attributes), or horizontal- or vertical partitioning. *Mappings* are executable expressions that specify how the data that conforms to one schema needs to be restructured to conform to another schema.

3. We have altered various model management operators in that the majority of them are defined to operate on *schematic correspondences* rather than *matches* as they are semantically richer than *matches*, and we have intro-duced additional operators for the automatic inference of *schematic corre-spondences* (`inferCorrespondence`) in addition to the operators `match` and `merge` for full support of the automatic initialisation or bootstrapping of a dataspace starting with the identification of *matches* followed by the infer-ence of *schematic correspondences* and the automatic generation of *mappings* [36]. To support the usage of the dataspace the operator `evaluateQuery`

has been added, which enables queries that are posed over any schema represented in the supermodel to be evaluated over multiple data sources using query unfolding [24].

4. In order to enable the incremental improvement of a dataspace, we have added the functionality to gather user feedback on query results, and to utilise the feedback provided to annotate the mappings with their precision and recall with respect to the feedback, to select the best mappings for future query executions and to refine the mappings (operators `annotateMappings`, `selectMappings`, and `refineMappings`) [4].

By building on top of the results of model management research and extending the model management operators with those listed above, we have created a toolkit that provides support for flexible management of schemas represented using heterogeneous models, enables the reaction to changes in the schemas of the underlying sources or the addition of new sources, provides flexibility with respect to the data models queries can be posed over and the ability to handle multiple integration schemas rather than just a single one. Furthermore, *DSToolkit* can support the whole dataspace lifecycle including initialisation, maintenance and improvement and it enables the casual user to provide feedback.

1.3   Contributions of the Paper

We present here *DSToolkit*, an architecture for flexible dataspace management that builds on the foundations provided by model management. The paper describes how the toolkit can be used to create dataspaces that exhibit various properties along the dimensions of dataspaces [26] rather than being limited to a specific point solution. The contributions of this paper are:

- Introduction of the *DSToolkit* and its comparison with prominent model management, data integration and dataspace management systems.
- Demonstration of its flexibility and ability to support the creation of dataspaces with different properties according to the dimensions of dataspaces using various examples.
- Illustration of the support that *DSToolkit* provides for the maintenance of a dataspace.
- Demonstration of the benefits that can be derived from user feedback, not just with respect to data sources already added to the dataspace, but also with respect to new data sources.

1.4   Organisation of the Paper

The paper is organised as follows: Section 2 presents a motivating example and gives a brief overview of the main functionality, capabilities and usage of *DSToolkit*. Section 3 places *DSToolkit* in the context of other prominent data integration and dataspace management systems, and discusses their commonalities and differences. Section 4 provides an overview of the architecture of the

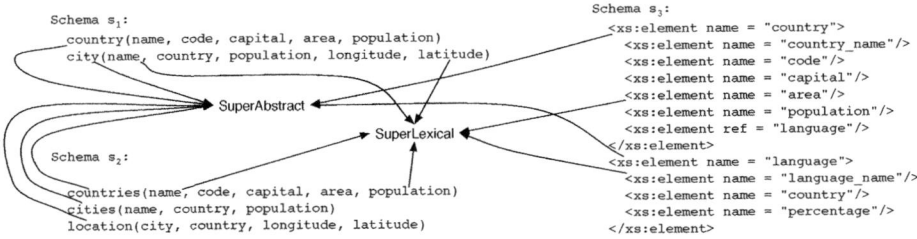

Fig. 1. Example source schemas

toolkit, followed by Section 5 which describes how *DSToolkit* supports the initialisation and shaping of various kinds of dataspaces and how it can be queried. Section 6 explains how the toolkit can be utilised to improve the dataspace and to evaluate complementary data sources with respect to already provided user feedback. Section 7 concludes the paper.

2   Motivating Example

This section introduces a motivating example and uses it to illustrate key features of *DSToolkit*, such as the flexibility provided by model management operators for managing multiple schemas and the associations between them, and the ability to utilise user feedback gathered on query result tuples to improve mappings over previously integrated sources.

As an example, we assume that the user would like to create a dataspace containing information on countries, their cities, the languages spoken in the countries, and that s/he has identified three data sources $d_1,...,d_3$ with the schemas $s_1,...,s_3$ shown in Figure 1. We further assume that d_1 contains information on european countries, d_2 information on african countries, and d_3 contains information on the languages spoken in all countries.

DSToolkit provides the user with a number of options to create a dataspace that meets the requirements, some of which are introduced in the following. After the import of the data sources into the system, *DSToolkit* offers multiple options for choosing or generating the preferred view of the data:

- Provide a manually defined global schema;
- Use the schema of any of the imported data sources;
- Use model management operators to generate a schema that meets the user's requirements.

As none of the three data sources in the example contains information on countries, their cities and their spoken languages, neither of the schemas of the imported sources can be utilised as global schema, leaving the other two options. One option would be to specify a schema manually that is to be used as a global schema and import it. However, as *DSToolkit* is based on model management, there is also the option to use the flexibility provided by the model management operators to generate a schema that conforms to the desired view of the data.

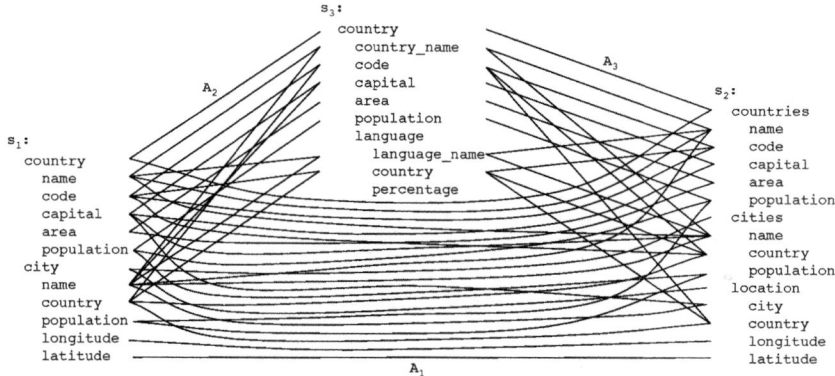

Fig. 2. Matches between source schemas

This can be achieved as follows: (i) Matching schemas pairwise with each other to identify the similarities, called *matches*, between them. Examples of matches between constructs in schemas s_1, s_2 and s_3 are shown in Figure 2. Some of the matches identified may be coincidental, e.g., the match between s_1.country. name and s_2.cities.name, which has been identified due to the same name of the two attributes. (ii) Using the information provided by the matches, semantically richer sets of *schematic correspondences* are inferred. Examples of schematic correspondences include **Same name, same construct (SNSC)**, which are represented in Figure 3 by lines between the corresponding constructs. Constructs can be attributes or tables in relational schemas or simple or complex elements in XML schemas. Other examples of schematic correspondences are **Different name same construct (DNSC)** (e.g., the attributes s_1.city.name and s_2.location.city represent the same constructs, i.e., the names of cities, but have different names), **Vertical partitioning (VP)** (e.g., the two tables **cities** and **location** in s_2 are a vertical partitioning of the table **city** in s_1), **Horizontal Partitioning (HP)**, and missing constructs, e.g., an attribute that is present in a table in one schema but not in the table that has been identified as corresponding in another schema. (iii) After the correspondences between schemas have been identified, they can be used to generate a merged integration schema.

The schema s_{m_2}, which is the schema generated by merging s_1, s_2 and s_3 is shown in Figure 4. The schematic correspondences between s_{m_2} and each of the source schemas are similar to those shown in Figure 3.

When the user is happy with the global schema created, the *mappings* between the global schema and each of the source schemas can be generated from the *schematic correspondences* between them [36]. The mappings for our example are shown in Figure 5.

After the mappings have been generated, the user can pose queries over the merged schema s_{m_2}. It is also possible to pose queries over any of the other schemas, e.g., source schemas or previously generated merged schemas, as long as the mappings between the schema and all the source schemas have been generated from schematic correspondences.

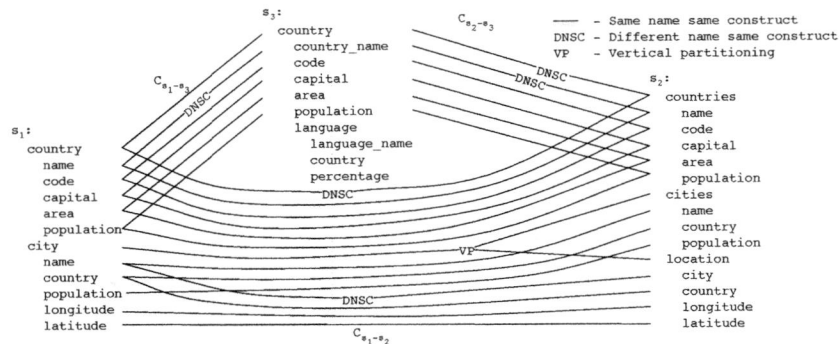

Fig. 3. Motivating example - schematic correspondences between source schemas

```
country(name, code, capital, area, population)
city(name, country, population, longitude, latitude)
language(country, name, percentage)
```

Fig. 4. Merged schema s_{m_2}

For example, let us assume that the user would like to get information on countries that are european and mediterranean, however, neither of the sources s/he has included so far in the dataspace contains information specific to mediterranean countries, though, as mentioned earlier d_1 contains information on european countries. Let us also assume that the user would like to run the query q_1 Select * from country o posed over s_{m_2}. As all three source schemas s_1, s_2 and s_3 contain information on countries, the mappings map_1, map_3 and map_5 are used to expand the query over the sources using query unfolding [24], which will result in european (from d_1), african (from d_2) and all other countries (from d_3) to be returned. To improve the initial integration, *DSToolkit* allows the user to annotate the query results indicating whether a result tuple was expected, i.e., is a true positive, or unexpected, i.e., is a false positive. We also allow the user to provide tuples that s/he expected to be returned as result but were not returned, i.e., false negatives. In our example, the user has annotated a small number of result tuples as shown in Table 1. The table also shows which mappings produced which of the annotated result tuples. Using the feedback, the mappings that produced the results are annotated with their precision and recall with respect to the user feedback provided.

$map_1 = <s_{m_2}$.country, select o.name, o.code, o.capital, o.area, o.population from s_1.country o>
$map_2 = <s_{m_2}$.city, select c.name, c.country, c.population, c.longitude, c.latitude from s_1.city c>
$map_3 = <s_{m_2}$.country, select o.name, o.code, o.capital, o.area, o.population from s_2.countries o>
$map_4 = <s_{m_2}$.city, select c.name, c.country, c.population, l.longitude, l.latitude
    from s_2.cities c, s_2.location l
    where c.name = l.city and c.country = l.country>
$map_5 = <s_{m_2}$.country, select o.country_name as name, o.code, o.capital, o.area, o.population
    from s_3.country o>
$map_6 = <s_{m_2}$.language, select l.country, l.language_name as name, l.percentage from s_3.country.language l>

Fig. 5. Mappings between s_{m_2} and s_1, s_2, s_3

Table 1. Annotated result tuples of q_1

name	code	capital	area	population	expected	not expected	mappings
France	F	Paris	547030	58317450	$\checkmark$		map_1, map_5
Turkey	TR	Ankara	780580	62484478	$\checkmark$		map_1, map_5
Italy	I	Rome	301230	57460274	$\checkmark$		map_1, map_5
Tunisia	TN	Tunis	163610	9019687		$\checkmark$	map_3, map_5
Morocco	MA	Rabat	446550	29779156		$\checkmark$	map_3, map_5
Algeria	DZ	Algier	2.38174e+06	29183032		$\checkmark$	map_3, map_5

Given existing feedback, the user may pose restriction on the mappings to be used when evaluating the query, e.g., to ensure that the proportion of the results that are true positives is high compared with those that are false positives. To do so, the user specifies a query together with requirements in terms of precision and recall that should be met by the query results returned. To achieve this, M' $\subseteq M$ are selected that are to be used to expand the query [4].

3   Related Work

This section discusses related work and compares it with *DSToolkit*. The work presented comes from several related areas, namely, traditional data integration, model management, dataspace management, evaluating the data quality of complementary data sources, and utilising feedback to support data integration.

3.1   On Traditional Data Integration

As data integration has been a research focus of the database community for several decades now, many contributions have been made. Here, we focus on two prominent representative examples, namely the IBM Information Server (IIS)[1], a commercial information integration platform [22], and Open II[2], an open source information integration suite [45].

The IBM Information Server provides a suite of tools that provide support for the various stages of information integration for both data materialisation or federation approaches [22]. However, even though a large number of tools are provided, they still require significant manual effort at various stages of the integration process, e.g., up-front to understand the data to be integrated, define an integration schema and data quality rules, or during the integration process to define or adjust mappings or to specify how to reconcile duplicates [22].

OpenII [45] provides an extensible platform for information integration consisting of a repository which uses a model-generic metamodel to represent schemas and mappings and a number of importers/exporters. *DSToolkit* also uses a model-generic metamodel to capture information on heterogeneous schemas and morphisms (matches, schematic correspondences, mappings). Open II provides a number of tools to aid several information integration tasks, e.g., *Harmony* for schema matching, visualising and debugging of matches identified

---

[1] www.ibm.com/software/data/infosphere/

[2] http://openii.sourceforge.net/

by multiple linguistic matchers, *Unity* to support the semi-automatic generation of mediated schemas, *RMap* and *XMap*, to generate the code needed for data exchange from matches identified by *Harmony* and confirmed or adjusted by the user.

As both *IIS* and *OpenII* provide support for traditional data integration they require a significant manual effort during the integration process and a good understanding of schemas and associations between them, thereby, making it almost impossible for the casual user to utilise either of them. In contrast, *DSToolkit* provides support for a fully automatic bootstrapping of the dataspace, i.e., the integration of various data sources, and only requires user interaction to provide feedback on result tuples, which can be provided by a casual user. Furthermore, neither IIS or OpenII are based on model management, even though *OpenII* makes use of a model-generic metamodel to represent schemas and matches, and therefore, are unable to benefit from the flexibility model management operators provide for handling multiple heterogeneous schemas and associations between their elements.

3.2   On Model Management

Model management [7,8] has been the focus of ongoing research for a number of years now and several systems have been proposed, e.g., Rondo [39], MISM [1], GeRoMe [31] and Automed [46]. All these systems use model-generic metamodels to abstract over the specifics of particular models and all provide importers at least for relational and XML schemas. The systems also provide a representation of morphisms between elements of the various schemas. The majority of the systems provide implementations for the model management operators `match`, which infers the morphisms between elements in different schemas, `merge`, which merges two schemas using the information provided in the morphisms between their elements, `modelGen`, which transforms a schema represented in one model into an equivalent schema represented in a different model, and some provide implementations of `compose`, which composes morphisms between schemas a, b and schemas b, c into morphisms between schemas a, c, and `difference` which returns the portion of a schema that does not participate in the morphisms between the schema and another schema.

Even though some of the systems have been around for a number of years and the purpose of being able to integrate multiple schemas is eventually to be able to query across their corresponding sources, only some of the model management platforms, namely GeRoMe [31] and Automed [37], have been extended for query answering. With the provision of model management operators, the majority of the platforms provide sufficient support for the maintenance in case of evolving schemas or additional sources to be integrated, however, even though it was speculated that model management platforms could provide the basis for dataspace management [8], the existing platforms tend not to have an analogue to `inferCorrespondence`, and have not been extended into a dataspace management platform that provides support for the 'pay-as-you-go' improvement that is characteristic of dataspace management.

3.3   On Dataspace Management

As dataspaces represent a fairly recent addition to the data integration landscape, the proposals have yet to reflect a shared understanding of best practice, and thus are diverse in their contributions across a variety of dimensions [26]. Proposals range from SEMEX [18], and iMeMex [15], both of which integrate personal information, over PayGo [35], which is targeted at the integration of web sources, to Q [48], the query system of ORCHESTRA [29], a collaborative data sharing system. None of the existing proposals for dataspaces are based on model management, making *DSToolkit* the first dataspace management system based on the solid foundations of model management and benefitting from the flexibility in managing diverse schemas provided by the approach. The majority of dataspace proposals tend to be point solutions addressing specific issues, but do not present a flexible approach that can be instantiated differently to create dataspaces with different properties along the dimensions identified in [26].

For example, SEMEX [18] requires an integration schema to be provided manually, whereas PayGo[35] forms a union schema, but neither approach provides support for generating a merged schema, as *DSToolkit* does in addition to accepting a manually specified integration schema or the option of choosing any of the source schemas as preferred view over the integrated data. SEMEX also provides no support for incremental improvement and even though PayGo[35] advocates incremental improvement, no details are provided on how this is achieved. In contrast, iMeMex starts with a union schema of all the integrated schema and provides support for manual improvement in the form of path-based queries called *trails*[15]. In contrast to our approach for incremental improvement which only requires users to indicate which result tuples meet their expectations, the need to provide path-based queries seems likely to exclude the casual user from improving the dataspace.

Similar to *DSToolkit*, UDI [14] can generate a merged schema automatically by matching source schemas and deriving a merged schema, but it makes simplifying assumptions in that the source schemas are limited to relational schemas with a single relation. UDI also provides no support for incremental improvement, which is the focus of Roomba [30]. Users are asked to provide feedback on matchings and mappings that have been determined by the system to provide the most benefit to the integration if annotated. Rather than requiring users to annotate matchings and mappings, Q [48], the query system of ORCHESTRA [29], asks users to provide feedback on the query results, similar to the feedback gathered by *DSToolkit*, and their rankings. This information is then propagated to the rankings of the matchings and mappings that produced the results.

3.4   On Evaluating the Data Quality of Complementary Data Sources

Data quality can be seen as how well the data meets the user's requirements, i.e., can be characterised as its "fitness for use"[49]. There are multiple measurable quality dimensions, e.g., accuracy, completeness, or currency [49].

A number of approaches have been proposed for quality-driven information integration (e.g., [42,44]). In [42] the authors propose an approach for quality-aware query plan creation and selection of the plan with the best weighted aggregate quality score according to several specific quality criteria.

In the Data Quality Broker, which is part of the DaQuinCIS architecture [44], queries posed over a global schema are unfolded into queries over multiple sources, an approach also followed in *DSToolkit*. The values of results returned by each source are annotated by the source with estimates of their data quality, e.g., their accuracy or completeness. The data quality dimensions are defined such that they can be measured, e.g., the accuracy of a value is measured in terms of its edit distance to values in reference dictionaries, and the completeness of a result tuple is measured in terms of the number of its attributes that are not null. This information is used to reconcile result tuples that refer to the same entities, but also to propagate the best information as determined by the quality values attached back to the sources which have returned data of lower quality to improve the overall data quality in the cooperative information system.

In contrast to both approaches, where information on the quality of the data is provided by the source providers themselves, in *DSToolkit* the quality of the information provided by a source is inferred from the user feedback provided on results of queries evaluated over those sources. The annotation of mappings with quality criteria in [42] could be compared to the annotation of mappings with their precision and recall with respect to the user feedback provided on the query results in *DSToolkit*, with the difference of the origin of the quality annotation. We could see precision as an indication for the accuracy and recall as an indication of the completeness of the results returned by a given mapping.

3.5   On Using Feedback to Support Data Integration

User feedback is a growing theme in data integration literature [5]. It is seen by many researchers as the key ingredient to face the difficulties that lie in the specification of data integration components. For example, Chiticariu *et al.*[12] proposed a method for generating integration schema. To ensure the suitability of the schema generated to user requirements, feedback is solicited from end-users. McCann *et al.* [38] developed a community-based system that solicits feedback from end users with the view to informing the schema matching operation. In doing so, the feedback is used to assess the matches between attributes in two schemas. Feedback has also been proposed as a means for driving the specification of schema mappings. For example Jeffery, *et al.* [30] developed a decision-theoretic framework for specifying the order in which candidate mappings can be confirmed by soliciting feedback from users with the objective of providing the *most benefit* to a dataspace. To do so, they developed a utility function that estimates the benefit that can be drawn from knowing whether a given schema mapping is correct or not.

User feedback has also been used as a means for authoring integration queries, i.e., queries that involve multiple data sources. The Q system supports such a functionality [48]. Specifically, given a set of keywords specified by the user,

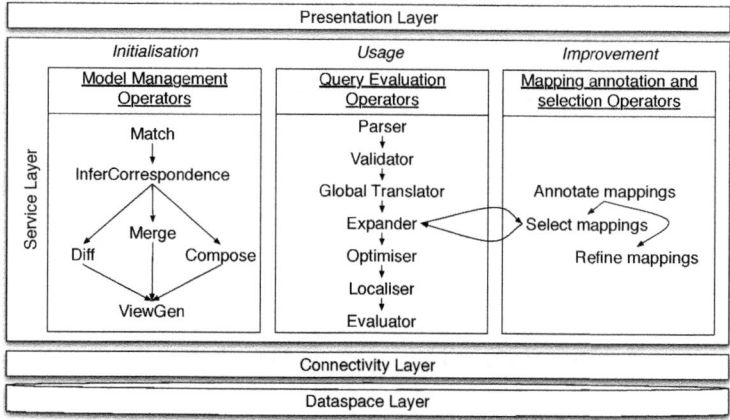

Fig. 6. Layered architecture of *DSToolkit*

the system suggests a list of candidate queries that may or may not meet user expectations. The user comments on the results returned by those queries. Based on these comments, the system ranks the list of candidate queries, the first query being the one that seems to meet user expectations the most.

While the above proposals show the key role user feedback can play to support data integration, they are confined to the use of feedback to support a single functionality. Differently, in our work, we try to make the most of the feedback supplied by end users, and use them to support three functionalities that are key to dataspaces improvement, viz., mapping annotation, selection and refinement.

4   The DSToolkit Architecture

This section provides an overview of the architecture of *DSToolkit*. It is implemented following the layered architecture shown in Figure 6. The arrows between the operators indicate in which order they tend to be utilised in each phase of the life cycle of a dataspace, namely, initialisation, usage and improvement. The arrows also indicate the dependency of the operators on the output of a previous operator producing the input of another operator, e.g., during initialisation the operator `inferCorrespondence` uses the output of `match`, i.e., the matches produced, as input to infer the schematic correspondences which, in turn are used by `merge` as input along with the schemas to be merged.

DSToolkit consists of a *dataspace* layer that persistently stores the model-generic representation of schemas, the morphisms between them, i.e., the matches, schematic correspondences supported by the matches, and the mappings derived from the latter, queries posed, the corresponding results as well as user feedback gathered on the results. The UML diagram of the model-generic representation of schemas is shown in Figure 7 and introduced in Section 5.1. The UML diagram of the morphism model is shown in Figure 8 and discussed in Section 5.2.

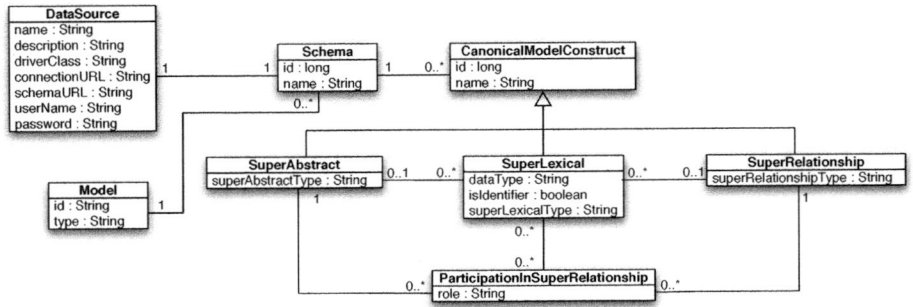

Fig. 7. UML Diagram of the canonical model

The *connectivity* layer provides the means for storing the information represented in the *dataspace* layer persistently and for accessing that information. As the functionality provided by this layer is fairly straightforward, it is omitted from further discussion. The *connectivity* layer is used in turn by the *service* layer which contains the actual functionality of *DSToolkit*, namely the model management operators, the query processor, and the techniques for mapping annotation, selection and refinement based on user feedback provided on result tuples. The functionality provided for use during the initialisation and usage phase is discussed in more detail in Section 5, whereas the functionality provided for use during the improvement phase is introduced in Section 6. The *presentation* layer exposes a web-based user interface through which the user can access the functionality provided by *DSToolkit*, i.e., the operators provided by the *service* layer and introduced in the following.

5   Initialising, Shaping and Querying a Dataspace with *DSToolkit*

This section describes how a dataspace is set-up. More specifically, it describes the initialisation of a dataspace, how data sources can be added, integration schemas inferred that match the user's view of the data and how queries can be posed and are evaluated. The section introduces the models required to capture all the necessary information, and the operators, some of which are model management operators, some of which we have added for the dataspace management. The operators provided by *DSToolkit* (except `addDataSource` and `evaluateQuery`) with their signatures and descriptions are listed in Table 3.

5.1   Model-Generic Canonical Model

This section presents the model-generic metamodel, called the *canonical model*, which is a generalisation of the MISM supermodel [1,2] and is shown in Figure 7. MISM consists of two levels of schema descriptions, namely, the model-specific description and the source-model independent supermodel. The model-specific

Table 2. Canonical model constructs, model-generic and model-specific constructs

Canonical model constructs	MISM model-generic constructs	Model-specific constructs per model			
		Relational	XSD	Object	Object Relational
Super-Abstract	*Abstract* *Aggregation* *StructOfAttributes*	Table	Root element Complex element	Class	Typed table Table Structured column
Super-Lexical	*Lexical*	Column	Simple element	Field	Column
Super-Relationship	*Foreign key* *Abstract attribute* *Generalisation*	Foreign key	Foreign key	Reference field Generalisation	Foreign key Reference Generalisation

descriptions contain all the constructs that are required to represent schemas in a particular model, e.g., table, column and foreign key for relational models and root element, complex element, simple element and foreign key for XSD. The supermodel contains a small set of model-generic constructs that represent the model-specific constructs by aggregating over their similarities, e.g., whether they can have values as is the case for relational and object-relational columns, simple elements in XSD and object fields, all of which correspond to lexical in the MISM model, or whether a construct represents referential integrity constraints between other constructs, such as the foreign key in XSD, relational and object-relational, all of which correspond to foreign key in MISM.

We have generalised the constructs of the MISM supermodel further according to whether the constructs can have values, represented by *SuperLexical*, e.g., column of a relational table, or a simple element in XSD, whether they are collection objects, represented by *SuperAbstract*, e.g., a relational table, or a complex element in XSD, or whether they represent a relationship between constructs, represented by *SuperRelationship*, e.g., a foreign key relationship in relational model or XSD, or the nesting of elements in XSD. Table 2 lists the constructs of our canonical model, the corresponding model-generic constructs of the MISM model and the corresponding model-specific constructs for relational, XSD, object and object-relational models. In addition to the example schemas shown in Figure 1, the figure also shows the corresponding construct in the canonical model, i.e., SuperAbstract for the relational tables in s_1 and s_2, as well as the root element (country) and the complex elements (language) in s_3. The relational columns in s_1 and s_2 and the simple elements in s_3 all correspond to SuperLexical, but for readability not all those correspondences are shown in Figure 1. Even though *DSToolkit* currently provides no importers for object and object-relational schemas (which can easily be added), as the *canonical model* is a generalisation of the MISM metamodel, providing support for object and object-relational schemas, our model supports both too. The exact type of each canonical model construct, e.g., whether a SuperAbstract represents a relational table or an XSD complex element or whether a SuperRelationship represents a relational or an XSD foreign key, is captured by the corresponding type (*superAbstractType, superLexicalType, superRelationshipType* in Fig. 7) as

Table 3. Operators provided by *DSToolkit*

Operation	Signature	Description
match	$A \leftarrow \mathtt{match}(s_i, s_j, [d_i, d_j])$	Match schemas s_i and s_j with each other using schema-based and if instances from data sources d_i and d_j with schemas s_i and s_j are provided, instance-based matchers to obtain set A of matches between them.
infer Correspondences	$C \leftarrow$ $\mathtt{inferCorrespondences}(A)$	Generate set of schematic correspondences C from set of matches A.
compose	$C \leftarrow \mathtt{compose}(C_1, C_2)$	Compose sets of schematic correspondences C_1, C_2 that hold between s_i, s_j and s_j, s_k into correspondences C that hold between s_i, s_k.
merge	$(s_m, C_1, C_2) \leftarrow$ $\mathtt{merge}(s_i, s_j, C)$	Merge two schemas s_i and s_j using the schematic correspondences C that hold between them and return merged schema s_m with correspondences C_1 and C_2 that hold between s_m, s_i and s_m, s_j.
difference	$C' \leftarrow \mathtt{difference}(s_i, s_j, C)$	Return schematic correspondences C' that represent differences between schemas s_i and s_j.
viewGen	$M \leftarrow \mathtt{viewGen}(C)$	Derive set of mappings M from set of schematic correspondences C.
selectMappings	$M' \leftarrow \mathtt{selectMappings}(M, \text{precisionTarget, recall-Target})$	Given a set of mappings M and a precision- or recall target λ select mappings $M' \subseteq M$ to be used to answer a query such that union of results returned by selected mappings M' achieve λ.
refineMappings	$M' \leftarrow \mathtt{refineMappings}(M)$	Given a set of mappings M produce mappings M' that meet user's requirements with respect to feedback provided on results better than M.

this information is later required for query rewriting (see Section 5.3). Super-Abstracts participating in a SuperRelationship can play different roles, e.g., the nested child element or its parent element, the referenced element in a foreign key relationship or the referencing element. Some relationships are also further specified by SuperLexicals, e.g., the attributes that form the (composite) foreign key. This information is captured in *ParticipationInSuperRelationship*.

When a *data source* d_i with its *schema* s_i is added using the operator (d_i, s_i) $\leftarrow$ addDataSource(*dataSourceName, description, driverClass, url, userName, password*) for data sources where the schema information can be obtained from the same url as the data or $(d_i, s_i) \leftarrow$ addDataSource(*dataSourceName, description, driverClass, url, schemaUrl, userName, password*) for data sources where the schema information can be found in a different location to the data source itself, the schema information is imported according to the correspondence between the model-specific constructs and the canonical model constructs. All the remaining operators operate over the representation of schemas in the canonical model, thereby abstracting over the differences of heterogeneous models.

5.2   The Bootstrapping Process and the Morphisms Generated

Here we present the bootstrapping process with the operators to generate the various morphisms expressing relationships between heterogeneous schemas and the models to capture them. The steps required to determine the morphisms between schemas, merge the schemas and generate mappings that specify how the data needs to be transformed are explained in more detail in the following.

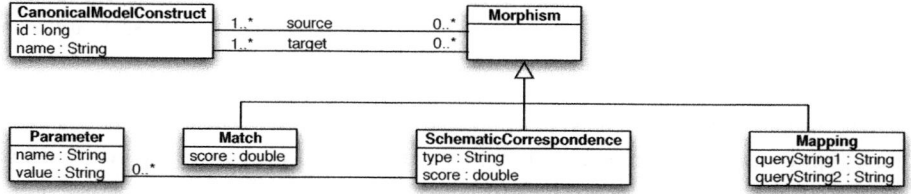

Fig. 8. UML Diagram of the morphism model

We also introduce multiple specialised kinds of morphisms, which were presented in Rondo [39]. A Morphism in general represents a binary relationship between two sets of instances of constructs. We distinguish between three kinds of morphisms of varying semantic richness, which we introduce in more detail in the following. A UML diagram of the different kinds of morphisms with their specific properties, their generalisation and their associations with *CanonicalModelConstructs* is shown in Figure 8.

Match and Infer Schematic Correspondences between Schemas. To be able to evaluate queries across multiple data sources, the relationships between the constructs in their schemas need to be identified and expressed in a way that can be utilised for unfolding queries posed over a schema of choice. To identify those relationships, firstly, the schemas need to be matched. A *match a* is a bidirectional morphism between sets of constructs which indicates that the constructs are similar to a certain extent; the confidence in which is indicated by the (similarity) *score* property of the *match a*, whereby a higher score indicates more confidence. A large corpus of literature is available on various matching approaches and algorithms (e.g., see [43,17]). To identify the set A of matches between two schemas s_i and s_j the operator $A \leftarrow \mathtt{match}(s_i, s_j, [d_i, d_j])$ can be used, which uses existing schema-, and if optional instance data is available from data sources d_i and d_j, instance-based matchers [16,41,3]. Examples of matchers include string based matchers using, e.g., edit distance or n-grams to determine how similar two strings are, data type matchers comparing the data types of constructs, or structure-based matchers comparing the structure of constructs within a schema such as the nesting of elements in XSD. The computational complexity of schema matchers is generally $O(n^2)$ [9].

However, these matches do not provide sufficient information for deriving mappings that express how data is to be transformed. To bridge the semantic gap between semantically poor matches and semantically rich mappings and to enable the automatic generation of mappings from matches, we use *schematic correspondences*, which are based on schematic heterogeneities introduced in [33,32] and which, as shown in [36], provide enough information to infer the mappings automatically. The operation $C \leftarrow \mathtt{inferCorrespondences}(A)$ is used to infer a set of semantically rich schematic correspondences C from the set of matches A. A schematic correspondence has a *score*, which as before for matches indicates the confidence associated with the correspondence. Only a subset of the matches A provided as input may result in schematic correspondences, i.e., only those matches that provide the most support for schematic correspondences of the following types: *different name for same construct (DNSC), same name for same construct (SNSC), missing constructs (MC), horizontal - (HP) and vertical partitioning (VP)*. For example, in Figure 2 matches between `city.population` in s_1 and `countries.population` in s_2 and vice versa have been identified. However, as can be seen in Figure 3 these have not resulted in a schematic correspondence between these elements, as other matches have provided more evidence for correspondences between `city` in s_1 and `cities` in s_2 as well as `country` in s_1 and `countries` in s_2 and their corresponding attributes rather than between the `population` attribute in `city` and `country`.

The type of a schematic correspondence is captured in its property *type*. Some correspondences require additional parameters, e.g., for VP the join predicates, which are captured in the *Parameters* that can be associated with a correspondence (Fig. 8). A genetic algorithm is used to search the space of all possible schematic correspondences supported by the input matches and find an optimal solution, the runtime of which can be controlled by various parameters, such as population size and number of generations [40].

Model Management Operations for Manipulating Schemas and Correspondences. Once schematic correspondences between constructs in schemas have been identified, those and the schemas can be manipulated using model management operators [7,8]. Model management operators have been shown to be useful for various scenarios that are of importance in data integration, such as schema integration and evolving schemas [6]. In addition to `match` *DSToolkit* provides implementations of $(s_m, C_1, C_2) \leftarrow$ `merge`(s_i, s_j, C), which merges two schemas s_i and s_j utilising the schematic correspondences C that hold between s_i and s_j and returns the merged schema s_m with sets of schematic correspondences C_1 and C_2 that hold between s_m, s_i and s_m, s_j, respectively, $C \leftarrow$ `compose`(C_1, C_2) which composes sets of schematic correspondences C_1, C_2 that hold between s_i, s_j and s_j, s_k, respectively, into schematic correspondences C that hold between s_i, s_k and $C' \leftarrow$ `difference`(s_i, s_j, C), which returns schematic correspondences C' that represent the differences between the two schemas s_i and s_j. Differences can include missing constructs or different names for the same constructs. The computational complexity of the operators is $O(n^2)$.

Model management operators provide flexibility for creating merged schemas that meet the user's requirements by merging multiple schemas and composing schematic correspondences, or to choose any of the source schemas and generate schematic correspondences between the selected schema(s) and all the other source schemas. For example, assume that the user who created the merged schema s_{m_2} shown in Figure 4 by matching all the source schemas with each other, inferring schematic correspondences between them and creating the merged schema s_{m_2} would also like to get information on the continents the countries are located in. The user has found another relational data source d_5 with the schema s_5 shown in Figure 9. As the user is aware that matching s_5 against the

```
Schema s₅:

continent(name, area)
encompasses(country, continent, percentage)
```
Fig. 9. Schema s_5

```
Matches between s₁ and s₅:

<{s₁.country.name}, {s₅.continent.name}, 0.5>
<{s₁.country.area}, {s₅.continent.area}, 0.8>
<{s₁.country}, {s₅.continent}, 0.4>
<{s₁.city.name}, {s₅.continent.name}, 0.5>
<{s₁.country.code}, {s₅.encompasses.country}, 0.55>
<{s₁.city.country}, {s₅.encompasses.country}, 0.67>

Schematic correspondences between s₁ and s₅:

<{s₁.country}, {s₅.continent}, different name same construct, 0.3>
<{s₁.country.name}, {s₅.continent.name}, same name same construct, 0.5>
<{s₁.country.area}, {s₅.continent.area}, same name same construct, 0.8>
<{s₁.country.code}, {s₅.continent}, missing attribute, 0.9>
                              ...
<{s₁.country.population}, {s₅.continent}, missing attribute, 0.9>
<{s₁.country.code}, {s₅.encompasses.country}, different name same construct, 0.55>
<{s₁.city.country}, {s₅.encompasses.country}, same name same construct, 0.67>
```

Fig. 10. Matches and schematic correspondences between s_1 and s_5

integration schema s_{m_2} might miss the association between s_{m_2}.country.code and s_5.encompasses.country due to the different names of the attributes and the lack of instances for s_{m_2} he decides to match s_5 with s_1 making use of instance data and infers the schematic correspondences between them, but he could have chosen any of the other source schemas. The results of $A_5 \leftarrow$ match(s_1, s_5, d_1, d_5) and $C_{s_1-s_5} \leftarrow$ inferCorrespondences(A_5) are shown in Figure 10.

To be able to merge s_{m_2} with s_5 the correspondences between the two schemas are needed, which can be obtained by composing the correspondences between s_{m_2}, s_1 and s_1, s_5: $C_{s_{m_2}-s_5} \leftarrow$ compose($C_{s_{m_2}-s_1}$, $C_{s_1-s_5}$). These correspondences are then used for merging s_{m_2} with s_5 to create s_{m_3}: (s_{m_3}, $C_{m_3-s_{m_2}}$, $C_{s_{m_3}-s_5}$) $\leftarrow$ merge(s_{m_2}, s_5, $C_{s_{m_2}-s_5}$). Compose is then used to generate the schematic correspondences between the newly merged schema s_5 and all the other source schemas by composing the correspondences between s_{m_3}, s_{m_2} and those between s_{m_2} and each of the source schemas s_1, s_2 and s_3, respectively. If the user decides that one of the source schemas is actually the preferred schema to pose queries over, s_5 could be matched with the remaining schemas s_2 and s_3 and the correspondences inferred from the matches returned. The resulting correspondences would be similar to those shown in Figure 10. Using the model management operators as illustrated in Section 2 and here, the user has generated a number of schemas and a number of sets of schematic correspondences.

ViewGen and the Resulting Mappings. As shown in [36] the schematic correspondences provide enough information to generate mappings automatically using the operator $M \leftarrow$ `viewGen`(C), which generates mappings M that correspond to the schematic correspondences C. The mappings are executable expressions that specify in form of two query strings, the specific properties of a mapping, how data that conforms to one schema has to be transformed to conform to another schema. The operator can be applied to any schematic correspondences between any two schemas (source- or merged integration schemas), allowing users to choose their favourite schema to pose queries over.

Iterating over the schematic correspondences the corresponding view between the participating source- and target-SuperAbstracts is generated. Depending on the cardinality of the participating SuperAbstracts, and more specifically on the kind of schematic correspondence, different approaches are used to generate the executable mapping. For example, in the case of one-to-one schematic correspondences, such as same name for same construct or different name for same construct, the view for populating the single target SuperAbstract from the single source SuperAbstract is generated with additional renaming applied in the case of the latter schematic correspondence. In the case of a one-to-many schematic correspondence, e.g., horizontal partitioning or vertical partitioning, the executable mapping for populating the single target SuperAbstract from the multiple source SuperAbstracts is generated by applying the union in the former case or by applying the join on the key attributes that are present in all vertically partitioned SuperAbstracts in the latter case. The computational complexity of the algorithm presented in [36] is $O(n)$.

5.3   Using the Dataspace: Evaluate Query

This section provides a brief overview of the expansion of queries posed over a schema represented in the canonical model into queries over potentially multiple sources, the translation of the source-specific sub-queries into the source-model-specific query languages and their evaluation [28].

We have defined *CMql*, a declarative query language inspired by SQL but defined over the constructs of the canonical model introduced in Section 5.1. A *CMql* query has the following form: SELECT $sl_1, ..., sl_n$ FROM $sa_1, ..., sa_m$ WHERE p, where $sl_1,...,sl_n$ is a project list of *SuperLexicals*, $sa_1,...,sa_m$ is a list of *SuperAbstracts*, and p is a conjunctive predicate. Queries can be posed over any schema, be it schemas of imported sources, global schemas generated using merge, or a manually specified global schema.

CMql queries are parsed, validated, translated into the algebra shown in Table 4 following standard translation schemes [20] during which selection predicates are pushed down into the SCAN operator, expanded, optimised, source-specific subqueries rewritten into the source-specific query languages and the query is evaluated with subqueries being sent to the query evaluator of the corresponding sources to be evaluated locally. The query processor consisting of components for each of those tasks (shown in Figure 6 - the components under Query Evaluation) is an extended version of the OGSA-DQP distributed query processor [34] which

Table 4. CMql algebra

Operator
SCAN(SuperAbstract, Predicate) → Collection
REDUCE(Collection, {SuperLexical}) → Collection
JOIN(Left_Collection, Right_Collection, Predicate) → Collection
UNION(Left_Collection, Right_Collection) → Collection
EvaluateSQL(dataSource , SQLqueryString, Predicate, {resultTuple})→ {resultTuple}
EvaluateXQuery(dataSource , XQueryString, Predicate, {resultTuple})→ {resultTuple}

has been adapted for the models used in *DSToolkit*. The UNION operator is used in the context of query unfolding whereas EvaluateSQL and EvaluateXQuery are used to evaluate the source-specific subqueries that have been rewritten into the source-model-specific query languages. Queries are expanded using query unfolding [24] with the mappings between the schema over which the query is posed and any of the schemas of the sources over which the query is to be evaluated. For expansion, either all the appropriate mappings generated by `viewGen` can be used or a subset of the mappings can be selected using the operator `selectMappings` which is explained in more detail in Section 6. The expanded queries are optimised, and subqueries of the optimised query plan that are associated with specific sources are translated into the source-specific query languages. The translated subqueries are passed to EvaluateSQL and EvaluateXQuery with information on the source over which the subquery is to be evaluated, i.e., which local query evaluator is to evaluate the subquery. Both operators can be parameterised with a predicate and result tuples, e.g., in the case of joins between two different sources. The query evaluator follows the iterator model [21], whereby each operator returns one result tuple at a time which can then be processed by the subsequent query operators, thereby removing the need to wait for a query operator to finish. A result tuple consists of multiple result values which in turn consist of a name of the corresponding superLexical and the actual value.

As mentioned earlier, *DSToolkit* not only allows multiple global schemas to co-exist, but also enables the user to pose queries over any schema of their choice, be it a merged schema, any of the source schemas or a manually provided schema that represents the user's preferred view of the data. For example, assume a user of data source d_1 would like to pose the following query q_3 over the dataspace:

```
SELECT c.name, c.population, c.longitude, c.latitude, o.name, o.population
FROM city c, country o
WHERE c.country = o.code
    AND o.population > 100000000
    AND c.population > 5000000.
```

However, s/he would like to pose the query over the schema s_1 of d_1 rather than any of the merged schemas, but would like the query to be evaluated not just over d_1, but also over d_2, which are the two data sources containing information on countries and their cities. The query is translated into the *CMql* algebra, expanded using query unfolding and the mappings between `city` and `country` in s_1 and s_2, which are similar to mappings map_1 and map_2 in Figure 5, optimised and the source-specific subqueries are rewritten. The

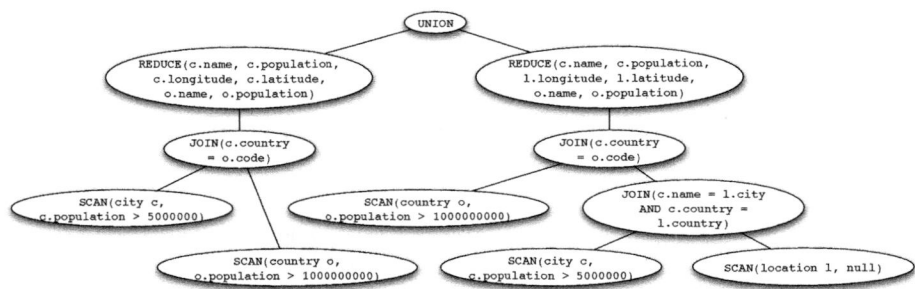

Fig. 11. Expanded query algebra tree of query q_3

```
UNION(
    EvaluateSQL(d_1,
        SELECT c.name, c.population, c.longitude, c.latitude, o.name, o.population
        FROM city c, country o
        WHERE c.country = o.code
            AND o.population > 100000000 AND c.population > 5000000, null, null),
    EvaluateSQL(d_2,
        SELECT c.name, c.population, l.longitude, l.latitude, o.name, o.population
        FROM city c, location l, country o
        WHERE c.name = l.city AND c.country = l.country
            AND c.country = o.code
            AND o.population > 100000000 AND c.population > 5000000, null, null))
```

Fig. 12. Expanded and rewritten query q_3

```
SELECT c.name, c.population, c.longitude, c.latitude, o.name, e.continent, l.name
FROM city c, country o, encompasses e, language l
WHERE c.country = o.code
    AND o.code = e.country
    AND o.code = l.country
    AND l.name = "spanish"
```

Fig. 13. Query q_4

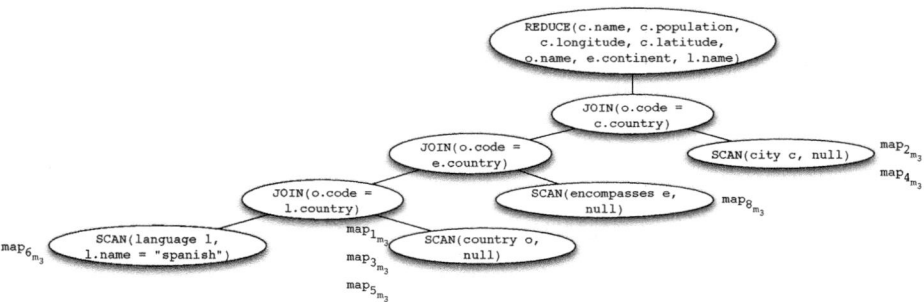

Fig. 14. Query algebra tree of query q_4

$map9_{m_3} = <s_{m_3}.\text{continent, select c.name, c.area from } s_5.\text{continent c}>$
$map10_{m_3} = <s_{m_3}.\text{encompasses, select e.country, e.continent, e.percentage from } s_5.\text{encompasses e}$

Fig. 15. Mappings between s_{m_3} and s_5

operator tree of the expanded query is shown in Figure 11 and the resulting query with the source-specific queries rewritten is shown in Figure 12.

We now assume that a different user of the same dataspace would like to pose the query q_4 shown in Figure 13 over the merged schema s_{m_3} and would like the query to be evaluated over all the sources that contain the relevant information on cities in countries in which spanish is spoken and the continent the country is located in. The query tree corresponding to q_3 is shown in Figure 14. In the figure the mappings that can be used to expand the query are also listed where mappings $map_{1_{m_3}}$, ..., $map_{6_{m_3}}$ are the mappings between s_{m_3} and each of the source schemas s_1, s_2, and s_3 and are comparable to the mappings with the corresponding names map_1, ..., map_6 shown in Figure 5. The remaining mappings between s_5 and s_{m_3} are shown in Figure 15.

Without any indication of which mappings should be used in cases where there are multiple mappings suitable to populate a particular construct, the query is expanded by creating the union of the query in which the constructs over which the query is posed are populated by each combination of the suitable mappings. As there are six different ways of combining available mappings the expanded query is rather large. For this reason, we only show the part of the query that has been expanded using mapping $map_{6_{m_3}}$ for language, $map_{1_{m_3}}$ for country, $map_{10_{m_3}}$ for encompasses and $map_{4_{m_3}}$ for city. The resulting part of the expanded and rewritten query is shown in Figure 16.

```
...
UNION(...,
    REDUCE(
        JOIN(
            JOIN(
                JOIN(
                    EvaluateXQuery(d_3,
                        <result>
                            let $s_3 := doc("...")
                            for $l in $s_3/country/language
                            where $l/name = "spanish"
                            return
                                <tuple>
                                    <l.country>{fn:data($l/country)}</l.country>
                                    <l.name>{fn:data($l/language_name)}</l.name>
                                    <l.percentage>{fn:data($l/percentage)}</l.percentage>
                                </tuple>
                        </result>, null, null),
                    EvaluateSQL(d_1,
                        SELECT o.name, o.code, o.capital, o.area, o.population
                        FROM country o, null, null),
                    o.code = l.country),
                EvaluateSQL(d_5,
                    SELECT e.country, e.continent, e.percentage
                    FROM encompasses e, null, null),
                o.code = e.country),
            EvaluateSQL(d_2,
                SELECT c.name, c.country, c.population, l.longitude, l.latitude
                FROM cities c, location l
                WHERE c.name = l.city AND c.country = l.country
                    AND l.name = "spanish", null, null),
            o.code = c.country),
        {c.name, c.population, c.longitude, c.latitude, o.name, e.continent, l.name}))
```

Fig. 16. Expanded and rewritten partial query q_4

6   Improving Dataspace Using User Feedback on Results

Without the knowledge of which mappings meet the user's requirements or intentions the closest, queries have to be expanded using the combination of all potential mappings. As illustrated in the previous section this can lead to rather complex queries that are evaluated over numerous sources, some of which may not contain any of the data the user is actually interested in. Also, not all combinations of mappings may return results that the user would like to see. However, rather than ask the user to provide feedback on the mappings themselves as in [11,10] or alter them to meet their needs, both of which requires a good understanding of schemas and mappings, thereby excluding the casual user, we ask the user to provide feedback on result tuples returned by the queries s/he has posed. This section describes the user feedback gathered and its application for annotating mappings, selecting and refining them to improve the integration over time [4]. This section also illustrates how previously gathered feedback can be utilised to evaluate the perceived quality of new complementary data sources with respect to the feedback previously provided, i.e., to evaluate how well a new data source matches the expectations of the user without requiring additional feedback from the user. The feedback can also be utilised along with new mappings over new sources for selection and refinement of mappings over already integrated sources. In contrast, rather than reusing previously provided feedback to gain information on new data sources, in [47] users need to provide feedback on the results returned by queries evaluated over the new sources.

Users can provide feedback on query result tuples of their own choosing indicating whether they expected a particular tuple that was returned to be present, i.e., a true positive (TP), or whether the tuple was not expected to be returned, i.e., a false positive (FP). Furthermore, users can also provide result tuples that were not returned, but that they expected to be part of the query result, i.e., false negatives (FN). It is worth mentioning that a true positive tuple of a given mapping may be a false negative tuple for another mapping. To illustrate this, consider that the user specifies that the tuple t is expected. The tuple t is a true positive for a given mapping m if m returns t, and is a false negative, otherwise. The feedback provides partial information on the extent of the construct in the schema over which the query was posed. This (partial) information can be used to calculate the precision and recall of the mappings that were used to produce those annotated results. Precision and recall are calculated as follows:

$$Precision(m, UF) = \frac{|TP(m, UF)|}{|TP(m, UF)| + |FP(m, UF)|} \tag{1}$$

$$Recall(m, UF) = \frac{|TP(m, UF)|}{|TP(m, UF)| + |FN(m, UF)|} \tag{2}$$

where $|TP(m, UF)|$, $|FP(m, UF)|$, $|FN(m, UF)|$ denote, respectively, the number of TPs, FPs and FNs returned by a mapping m according to the user feedback UF on query results involving the mapping m.

Table 5. Precision and recall of map_1, map_3 and map_5 based on user feedback in Table 1

Mapping	Precision	Recall
map_1	1	1
map_3	0	0
map_5	0.5	1

For example, returning to the user feedback provided on result tuples to query q_1 shown earlier in Table 1, this user feedback is utilised to annotate the mappings map_1, map_3 and map_5 with their respective precision and recall computed by Equations 1 and 2, respectively. The results are shown in Table 5. Another option for annotating the mappings would be the use of gold-standard ground truth data, if available, rather than asking the user for feedback. This option, however, is not discussed here.

6.1   Improving Mappings and Their Selection for Query Re-runs

The user feedback is not only used to annotate the mappings with their respective precision and recall based on the user feedback gathered. It can also be used to select the mappings that meet the user's requirements for future re-runs of queries and to generate better mappings with respect to the user's expectations using refinement. Both are explained in more detail in the following.

Mapping Selection. Using all candidate mappings to answer a user query may have undesirable consequences. In particular, there is a risk that the query will take a long time to be evaluated, if the number of candidate mappings is large, and that the resulting tuples are largely false positives, if the mappings are of poor quality. To overcome this issue, we implemented an operation that selects the mappings to be used to evaluate a user query. Not all users have the same requirements as to the completeness and soundness of the results. Because of this, the selection operation we developed provides the user with a means to specify his requirements with respect to results soundness, by specifying a threshold for precision, or to the completeness, by specifying a recall threshold. Specifically, given a set of candidate mappings M and a precision- or recall target λ set by the user, the $M' \leftarrow \texttt{selectMappings}(M, \text{precisionTarget}, \text{recallTarget})$ operator selects the mappings $M' \subseteq M$ that are to be used to answer the query such that the union of query results returned by the selected mappings M' achieve λ. The problem is formulated as a constrained optimisation problem in which the selected target λ (say precision) is the constraint and the other value (say recall) is maximised and is solved using tabu search [40].

For an example, we revisit the user feedback provided on results of query q_1 in Table 1 and the resulting precision and recall of the corresponding mappings shown in Table 5. If the user chooses a desired precision target $0.5 < \lambda \leq 1$, only mapping map_1 would be used to answer the query, whereas for a precision target $0 < \lambda \leq 0.5$ both mappings map_1 and map_5 would be used. The recall of the mappings is either 1 or 0, so any recall target of $0 < \lambda \leq 1$ would exclude mapping map_3 from being utilised.

Mapping selection cannot only be used to exclude mappings that are correct in the sense that they associate the correct constructs with each other but do not meet the user's requirements with respect to the results they return, e.g., all three mappings map_1, map_3 and map_3 associate the construct representing countries in s_{m_2} with the constructs representing countries in the corresponding source schemas, however, not all the results returned meet the (subjective) expectation of the user. It can also be used to exclude incorrect mappings, which could be produced by the automatic bootstrapping process. For example, lets assume we have an additional data source d_6 with the following schema s_6:

```
province(name, country, capital, area, population)
```

Using the automatic bootstrapping process, the following incorrect mapping map_{11} could have been produced, associating incorrectly the construct representing countries in s_{m_3} with the construct representing province in s_6:

$map_{11} = <s_{m_3}$.country, select p.name, p.country as code, p.capital, p.area, p.population from s_6.province p>

which would return information on provinces rather than countries. After the user has annotated some results returned by this mapping as false positives and chosen a precision target of $0 < \lambda$ this mapping is excluded from further query expansions. Gathering feedback only on result tuples which contain the data the user is familiar with rather than mappings enables casual users to improve the dataspace themselves rather than having to rely on developers who have a good understanding of both schemas and mappings.

Mapping Refinement. Mapping selection can only be successful in returning exactly the query results the user expects if the mappings meet the user's requirements. However, this may not be the case, e.g., as mentioned earlier d_1 contains information on european and d_2 information on african cities and countries, i.e., they form a horizontal partitioning. This, however, was not detected during the bootstrapping process as can be seen in the lack of a mapping that creates the union of the information from both sources (Fig. 5). If a user is interested in information from both european and african cities and countries, neither of the mappings between the merged schema and each of the source schemas will fully meet the requirements, only a union would achieve this.

The operator $M' \leftarrow$ refineMappings(M) aims to produce mappings M' that meet the user's requirements with respect to the feedback provided on result tuples better than the existing mappings M by trying to increase the number of true positives and/or reduce the number of false positives [4]. False positives can be reduced by filtering the results using the operators of the relational algebra that allow filtering, namely, join, intersection and difference. To increase the number of true positives union can be used. The space of mappings that can

$map_{12} = <s_{m_2}.country,$
   select name, code, capital, area, population
   from (
      (select o.name, o.code, o.capital, o.area, o.population
      from s_1.country o)
   union
      (select o.name, o.code, o.capital, o.area, o.population
      from s_2.countries o)
)>

Fig. 17. Mapping map_{12} created by refinement

be generated by creating the join, intersection, difference or union of existing mappings M is very large. To explore the space, an evolutionary algorithm [40] is used which creates new mappings from existing mappings by mutating mappings, i.e., applying join, or combining mappings by applying intersection, difference or union (for more detail on the approach, see [4]).

Consider for example a user who is interested in information about european and african countries, and consider the following two candidate mappings:

$map_1 = <s_{m_2}.country,$ select o.name, o.code, o.capital, o.area, o.population from s_1.country o>
$map_3 = <s_{m_2}.country,$ select o.name, o.code, o.capital, o.area, o.population from s_2.countries o>

Both these mappings return tuples that are of relevance to the user. The two mappings, however, do not return the same set of expected tuples. This suggests an opportunity for increasing the recall by unioning the source queries of the two mappings. Using our refinement algorithm, we were able to create a new mapping map_{12} shown in Figure 17 to increase the recall of the results.

6.2   Assessing the Quality of New Data Sources and Improving the Mappings over Existing Data Sources

As well as applying mapping annotation, selection and refinement to mappings over already integrated data sources that have been used to expand queries the user has posed and the results of which the user has annotated, these techniques can also be used for mappings over new complementary data sources without requiring additional feedback. When a new data source is integrated such that mappings are generated that populate constructs in an existing global schema from the new source, previously executed queries can be re-run automatically and the tuples returned by the new mappings compared with tuples previously annotated by the user. When an annotated tuple with the same attribute values as a result tuple returned from the new source is found, this annotation is transferred to the new result tuple, thereby annotating tuples that have been produced using new mappings over the new source without requiring additional feedback. Once the result tuples returned by the new mappings are annotated, this information can be used to annotate those new mappings using the process described earlier in this section. This gives an indication of how well the new source meets the user's requirements with respect to the results its corresponding mappings return and with respect to the user feedback gathered previously.

Table 6. Annotated result tuples of q_5

name	code	capital	area	population	expected	not expected	mappings
Iceland	IS	Reykjavik	103000	270292	$\checkmark$		map_1
United Kingdom	GB	London	244820	58489975	$\checkmark$		map_1
Belarus	BY	Minsk	207600	10415973		$\checkmark$	map_1
Liechtenstein	FL	Vaduz	160	31122		$\checkmark$	map_1
Germany	D	Berlin	356910	83536115		$\checkmark$	map_1

For an example, we revisit the query q_1 of the user who is only interested in countries that are both european and mediterranean and has provided the feedback on some of the result tuples as shown in Table 1 indicating his expectations. Lets further assume that the user added the new data source d_4 containing information on mediterranean countries with the schema s_4:

```
mediterraneanCountry(name, code, capital, area, population)
```

and that the mapping map_7 was generated:

$map_7 = <s_{m_2}$.country,
   select m.name, m.code, m.capital, m.area, m.population
   from s_4.mediterraneanCountry m>

A rerun of query q_1 over d_4 using map_7 to expand the query will return amongst other tuples all the tuples shown in Table 1 as all the countries listed in the table and annotated by the user are mediterranean countries. The previously provided feedback shown in the table is used to annotate map_7 with its precision of 0.5 and its recall of 1.

As soon as mappings are annotated with their respective precision and recall this information can be utilised for mapping selection and refinement. This can be done independent of whether the annotation is based on user feedback provided on result tuples or whether the annotation has been inferred automatically by comparing tuples produced by new mappings with those that the user annotated previously.

As an example for selection of new mappings based on their automatically inferred annotation, we consider the mapping map_7 with its inferred precision and recall that are better than those of map_3. This new mapping will be selected along with map_1 and map_5 for future evaluations of q_1 as long as a precision target λ of $0 < \lambda \leq 0.5$ or a recall target of $0 < \lambda \leq 1$ is specified.

As an example for the refinement of existing mappings using information from new sources, let us assume that a user is interested in european countries that are located on an island. The user has previously executed the query q_5 SELECT * FROM country o posed over s_{m_2} but stipulated that the query should only be evaluated over d_1, as this source contains only information on european countries, which results in a large number of unexpected results as the majority of european countries are not located on an island. The user previously provided the feedback on a small number of result tuples shown in Table 6.

He has managed to find a data source d_7 containing information on countries that are located on islands and information on that island with the schemas s_7:

```
locatedOnIsland(country, island_name, area, longitude, latitude)
```

During the bootstrapping process the new information has been added to s_{m_2}, resulting in this schema:

```
country(name, code, capital, area, population)
city(name, country, population, longitude, latitude)
language(country, name, percentage)
locatedOnIsland(country, island_name, area, longitude, latitude)
```

with the following mapping to populate `locatedOnIsland` in the new merged schema with the information in d_7:

$map_{13} = <s_{m_2}$.locatedOnIsland,
     select l.country, l.island_name, l.area, l.longitude, l.latitude
     from s_7.locatedOnIsland l>

Adding the information from s_7 has not resulted in new mappings for information that was already part of the schema, e.g., `country`, but during the bootstrapping process it has been identified by `match` that `locatedOnIsland.country` represents the same information as `country.code`, namely the abbreviated names of countries. Using this information and the previously gathered feedback shown in Table 6 the refinement algorithm creates the following mapping, which joins the information on countries with the information on which countries are located on islands, thereby reducing the number of false positives, i.e., the european countries that are not located on islands.

$map_{14} = <s_{m_2}$.country,
     select o.name, o.code, o.capital, o.area, o.population
     from s_1.country o, s_7.locatedOnIsland l
     where o.code = l.country>

Rerunning query q_5 using the new mapping map_{14} to expand the query returns information on all the countries that are european and are located on an island, i.e., all the tuples that are expected by the user but none of the tuples that are not expected as indicated by the feedback in Table 6. This means that map_{14} is annotated with its precision of 1 and its recall of 1. The mapping map_{13}, which provides completely new information on islands, cannot be annotated, though, as the tuples it produces have not been returned by any previously run query and, therefore, have not been annotated with feedback that can be reused.

7 Conclusions

In this paper, we presented *DSToolkit*, an architecture for flexible dataspace management and the first dataspace management system that:

1. Supports the whole lifecycle of a dataspace, namely initialisation, maintenance, usage and improvement. For usage, i.e., querying across multiple data sources, *DSToolkit* provides the means for structured *CMql* queries posed over any schema to be expanded, optimised and for source-specific sub-queries to be translated into the source-model-specific query language for evaluation over the data sources, thereby meeting one of the requirements for a dataspace management systems [19] and sub-challenge 2.2 in [25]. *DSToolkit* enables the user to specify the precision- or recall-target that the query results should meet, which is part of the requirement identified as sub-challenge 4.1 in [25]. However, *DSToolkit* does not provide support for keyword queries yet nor does it deal with uncertain or inconsistent data.

2. Is based on model management, thereby benefitting from the flexibility provided by the model management operators for managing heterogeneous models and associations between them. In the dataspace vision, it was highlighted that a dataspace management system needs to be able to support multiple data models and cope with integrated data sources over which it has no full control [19,25]. Building on the basis of model management means that *DSToolkit* provides support for both, dealing with different data models and changes in the schemas of the integrated data sources. However, *DSToolkit* only provides support for structured data sources, but not unstructured.

3. Enables casual users to improve the integration by providing feedback on result tuples which is then utilised to annotate, select and refine the mappings used to expand the queries. *DSToolkit* benefits from the interaction with the user, i.e., the feedback provided by the user, by utilising it to determine which mappings meet the users' requirements better than others and generating new mappings with the aim to improve the integration of the data sources. The need for the analysis of the users' interaction with the dataspace and the creation of additional relationships between sources or other forms of improvements of the dataspace was identified as challenge 5 in [25]. *DSToolkit* also estimates the quality of the query result in terms of its precision and recall based on previously gathered user feedback, a point identified as part of sub-challenge 2.2 in [25].

4. Can utilise the feedback gathered previously to determine the perceived quality of new data sources with respect to how well the data in the new source matches the user's expectations, to select and refine both mappings over previously integrated source and those over the new sources without requiring additional feedback from the user. In [25] the authors argue that it is important to reuse the information provided by users as much as possible (sub-challenge 5.3). *DSToolkit* reuses previously provided feedback for the integration of new data sources, therefore, reducing the amount of feedback the user has to provide.

The presented approach where the data remains in the sources, however, is not the only option for a dataspace system. The integrated data sets could also be curated, annotated and/or cleaned, which requires a (modified) copy of the data sources.

References

1. Atzeni, P., Bellomarini, L., Bugiotti, F., Gianforme, G.: Mism: A platform for model-independent solutions to model management problems. J. Data Semantics 14, 133–161 (2009)
2. Atzeni, P., Gianforme, G., Cappellari, P.: A universal metamodel and its dictionary. T. Large-Scale Data- and Knowledge-Centered Systems 1, 38–62 (2009)
3. Aumueller, D., Do, H.H., Massmann, S., Rahm, E.: Schema and ontology matching with coma++. In: SIGMOD Conference, pp. 906–908 (2005)
4. Belhajjame, K., Paton, N.W., Embury, S.M., Fernandes, A.A.A., Hedeler, C.: Feedback-based annotation, selection and refinement of schema mappings for dataspaces. In: EDBT, pp. 573–584 (2010)
5. Belhajjame, K., Paton, N.W., Fernandes, A.A.A., Hedeler, C., Embury, S.M.: User feedback as a first class citizen in information integration systems. In: CIDR, pp. 175–183 (2011)
6. Bernstein, P.A.: Applying model management to classical meta data problems. In: CIDR, pp. 209–220 (2003)
7. Bernstein, P.A., Halevy, A.Y., Pottinger, R.A.: A vision for management of complex models. SIGMOD Record 29(4), 55–63 (2000)
8. Bernstein, P.A., Melnik, S.: Model management 2.0: manipulating richer mappings. In: SIGMOD Conference, pp. 1–12 (2007)
9. Bernstein, P.A., Melnik, S., Petropoulos, M., Quix, C.: Industrial-strength schema matching. SIGMOD Record 33(4), 38–43 (2004)
10. Cao, H., Qi, Y., Candan, K.S., Sapino, M.L.: Feedback-driven result ranking and query refinement for exploring semi-structured data collections. In: EDBT, pp. 3–14 (2010)
11. Chai, X., Vuong, B.Q., Doan, A., Naughton, J.F.: Efficiently incorporating user feedback into information extraction and integration programs. In: SIGMOD Conference, pp. 87–100 (2009)
12. Chiticariu, L., Kolaitis, P.G., Popa, L.: Interactive generation of integrated schemas. In: SIGMOD Conference, pp. 833–846 (2008)
13. Chiticariu, L., Tan, W.C.: Debugging schema mappings with routes. In: VLDB, pp. 79–90 (2006)
14. Das Sarma, A., Dong, X., Halevy, A.: Bootstrapping pay-as-you-go data integration systems. In: SIGMOD, pp. 861–874 (2008)
15. Dittrich, J., Salles, M.A.V., Blunschi, L.: imemex: From search to information integration and back. IEEE Data Eng. Bull. 32(2), 28–35 (2009)
16. Do, H.H., Rahm, E.: Coma: a system for flexible combination of schema matching approaches. In: VLDB, pp. 610–621 (2002)
17. Do, H.H., Rahm, E.: Matching large schemas: Approaches and evaluation. Inf. Syst. 32(6), 857–885 (2007)
18. Dong, X., Halevy, A.Y.: A platform for personal information management and integration. In: CIDR, pp. 119–130 (2005)
19. Franklin, M.J., Halevy, A.Y., Maier, D.: From databases to dataspaces: a new abstraction for information management. SIGMOD Record 34(4), 27–33 (2005)
20. Garcia-Molina, H., Ullman, J.D., Widom, J.: Database Systems The Complete Book. Pearson International edn., 2nd edn. (2009)
21. Graefe, G.: Encapsulation of parallelism in the volcano query processing system. In: SIGMOD Conference, pp. 102–111 (1990)

22. Haas, L.: Beauty and the Beast: The Theory and Practice of Information Integration. In: Schwentick, T., Suciu, D. (eds.) ICDT 2007. LNCS, vol. 4353, pp. 28–43. Springer, Heidelberg (2006)
23. Haas, L., Lin, E., Roth, M.: Data integration through database federation. IBM Systems Journal 41(4), 578–596 (2002)
24. Halevy, A.Y.: Answering queries using views: A survey. The VLDB Journal 10(4), 270–294 (2001)
25. Halevy, A.Y., Franklin, M.J., Maier, D.: Principles of dataspace systems. In: PODS, pp. 1–9 (2006)
26. Hedeler, C., Belhajjame, K., Fernandes, A.A.A., Embury, S.M., Paton, N.W.: Dimensions of Dataspaces. In: Sexton, A.P. (ed.) BNCOD 2009. LNCS, vol. 5588, pp. 55–66. Springer, Heidelberg (2009)
27. Hedeler, C., Belhajjame, K., Paton, N.W., Campi, A., Fernandes, A.A.A., Embury, S.M.: Dataspaces. In: SeCO Workshop, pp. 114–134 (2009)
28. Hedeler, C., Paton, N.W.: Utilising the MISM Model Independent Schema Management Platform for Query Evaluation. In: Fernandes, A.A.A., Gray, A.J.G., Belhajjame, K. (eds.) BNCOD 2011. LNCS, vol. 7051, pp. 108–117. Springer, Heidelberg (2011)
29. Ives, Z.G., Green, T.J., Karvounarakis, G., Taylor, N.E., Tannen, V., Talukdar, P.P., Jacob, M., Pereira, F.: The orchestra collaborative data sharing system. SIGMOD Record 37(3), 26–32 (2008)
30. Jeffery, S.R., Franklin, M.J., Halevy, A.Y.: Pay-as-you-go user feedback for dataspace systems. In: SIGMOD Conference, pp. 847–860 (2008)
31. Kensche, D., Quix, C., Li, X., Li, Y., Jarke, M.: Generic schema mappings for composition and query answering. Data & Knowledge Engineering (DKE) 68(7), 599–621 (2009)
32. Kim, W., Choi, I., Gala, S.K., Scheevel, M.: On resolving schematic heterogeneity in multidatabase systems. Distributed and Parallel Databases 1(3), 251–279 (1993)
33. Kim, W., Seo, J.: Classifying schematic and data heterogeneity in multidatabase systems. IEEE Computer 24(12), 12–18 (1991)
34. Lynden, S., Mukherjee, A., Hume, A.C., Fernandes, A.A.A., Paton, N.W., Sakellariou, R., Watson, P.: The design and implementation of OGSA-DQP: A service-based distributed query processor. Future Generation Comp. Syst. 25(3), 224–236 (2009)
35. Madhavan, J., Cohen, S., Dong, X.L., Halevy, A.Y., Jeffery, S.R., Ko, D., Yu, C.: Web-scale data integration: You can afford to pay as you go. In: CIDR, pp. 342–350 (2007)
36. Mao, L., Belhajjame, K., Paton, N.W., Fernandes, A.A.A.: Defining and Using Schematic Correspondences for Automatically Generating Schema Mappings. In: van Eck, P., Gordijn, J., Wieringa, R. (eds.) CAiSE 2009. LNCS, vol. 5565, pp. 79–93. Springer, Heidelberg (2009)
37. McBrien, P., Poulovassilis, A.: P2P Query Reformulation over Both-As-View Data Transformation Rules. In: Moro, G., Bergamaschi, S., Joseph, S., Morin, J.-H., Ouksel, A.M. (eds.) DBISP2P 2005 and DBISP2P 2006. LNCS, vol. 4125, pp. 310–322. Springer, Heidelberg (2007)
38. McCann, R., Kramnik, A., Shen, W., Varadarajan, V., Sobulo, O., Doan, A.: Integrating data from disparate sources: A mass collaboration approach. In: ICDE, pp. 487–488 (2005)
39. Melnik, S., Rahm, E., Bernstein, P.A.: Rondo: a programming platform for generic model management. In: SIGMOD, pp. 193–204 (2003)

40. Michalewicz, Z., Fogel, D.: How to solve it: modern heuristics. Springer, Heidelberg (2000)
41. Mork, P., Seligman, L., Rosenthal, A., Korb, J., Wolf, C.: The harmony integration workbench. J. Data Semantics 11, 65–93 (2008)
42. Naumann, F., Leser, U., Freytag, J.C.: Quality-driven integration of heterogenous information systems. In: VLDB, pp. 447–458 (1999)
43. Rahm, E., Bernstein, P.A.: A survey of approaches to automatic schema matching. VLDB Journal 10(4), 334–350 (2001)
44. Scannapieco, M., Virgillito, A., Marchetti, C., Mecella, M., Baldoni, R.: The architecture: a platform for exchanging and improving data quality in cooperative information systems. Inf. Syst. 29(7), 551–582 (2004)
45. Seligman, L., Mork, P., Halevy, A.Y., Smith, K.P., Carey, M.J., Chen, K., Wolf, C., Madhavan, J., Kannan, A., Burdick, D.: Openii: an open source information integration toolkit. In: SIGMOD Conference, pp. 1057–1060 (2010)
46. Smith, A., Rizopoulos, N., McBrien, P.: AutoMed Model Management. In: Li, Q., Spaccapietra, S., Yu, E., Olivé, A. (eds.) ER 2008. LNCS, vol. 5231, pp. 542–543. Springer, Heidelberg (2008)
47. Talukdar, P.P., Ives, Z.G., Pereira, F.: Automatically incorporating new sources in keyword search-based data integration. In: SIGMOD Conference, pp. 387–398 (2010)
48. Talukdar, P.P., Jacob, M., Mehmood, M.S., Crammer, K., Ives, Z.G., Pereira, F., Guha, S.: Learning to create data-integrating queries. PVLDB 1(1), 785–796 (2008)
49. Wang, R.Y.: A product perspective on total data quality management. Commun. ACM 41(2), 58–65 (1998)

Temporal Content Management
and Website Modeling:
Putting Them Together

Paolo Atzeni, Pierluigi Del Nostro, and Stefano Paolozzi

Dipartimento di Informatica ed Automazione
Università degli Studi Roma Tre
Via della Vasca Navale, 79 – Rome, Italy
{atzeni,pdn,paolozzi}@dia.uniroma3.it

Abstract. The adoption of high-level models for temporal, data-intensive Web sites is proposed together with a methodology for the design and development through a content management system (CMS). The process starts with a traditional ER scheme; the various steps lead to a temporal ER scheme, to a navigation scheme (called N-ER) and finally to a logical scheme (called T-ADM). The logical model allows the definition of page-schemes with temporal aspects (which could be related to the page as a whole or to individual components of it). Each model considers the temporal features that are relevant at the respective level. A content management tool associated with the methodology has been developed: from a typical content management interface it automatically generates both the relational database (with the temporal features needed) supporting the site and the actual Web pages, which can be dynamic (JSP) or static (plain HTML or XML), or a combination thereof. The tool also includes other typical features of content management all integrated with temporal features.

1   Introduction

The systematic development of Web contents has attracted the interest of the database community as soon as it was realized that the Web could be used as a suitable means for the publication of useful information for communities of users (Atzeni et al. [1], Ceri et al. [2], Fernández et al. [3], Brambilla et al. [16]). Specific attention has been devoted to *data-intensive* sites, where the information of interest has both a somehow regular structure and a possibly significant volume; here the information can be profitably stored as data in a database and the sites can be generated (statically or dynamically) by means of suitable expressions (i.e. queries) over them (Merialdo et al. [4]). In this scenario, the usefulness of high-level models for the intensional description of Web sites has been advocated by various authors, including Atzeni et al. [1, 4] and Ceri et al. [2, 17], which both propose logical models in a sort of traditional database sense and a *model-based development* for data intensive Web sites.

A. Hameurlain et al. (Eds.): TLDKS V, LNCS 7100, pp. 158–182, 2012.
© Springer-Verlag Berlin Heidelberg 2012

When accessing a Web site, users would often get significant benefit from the availability of time-related information, in various forms: from the history of data in pages to the date of the last update of a page (or the date the content of a page was last validated), from the access to previous versions of a page to the navigation over a site at a specific past date (with links coherent with respect to this date). As common experience tells, various aspects of a Web site often change over time: (i) *the actual content of data* (for example, in a University Web site, the instructor for a course); (ii) *the types of data offered* (at some point we could decide to publish not only the instructor, but also the teaching assistants, TAs, for a course); (iii) *the hypertext structure* (we could have the instructors in a list for all courses and the TAs only in separate detail pages, and then change, in order to have also the TAs in the summary page); (iv) *the presentation* (we could present different pieces of information on the basis of the particular visitor of the Web site). Indeed, most current sites do handle very little time-related information, with past versions not (or only partially) available and histories difficult to reconstruct, even when there is past data. Clearly, these issues correspond to cases that occur often, with similar needs, and that could be properly handled by specific techniques for supporting time-related aspects. We believe that a general approach to this problem could generate significant benefits to many data-intensive Web applications.

Indeed, we have here requirements that are analogous to those that led to the development of techniques for the effective support to the management of time in databases by means of *temporal database* (see [5], [7], [18] and [19] for interesting discussions).

It is well known that in temporal databases there are various dimensions along which time can be considered. Beside *user-defined time* (the semantics of which is "known only to the user", and therefore is not explicitly handled), there are *valid time* ("the time a fact was true in reality") and *transaction time* ("the time the fact was stored in the database").

The different temporal dimensions are orthogonal and are usually implemented with two database columns, corresponding with the bounds of the temporal period. The difference between the temporal dimensions is in the semantic of the periods and consequently how they are managed. Let us consider the two main temporal dimensions: transaction time and validity time. Transaction time is usually managed exclusively by the system and should never be modified by the user. The (closed) lower bound of a transaction time period represents the time a transaction inserted the data in the database and the (open) upper bound the time data has been deleted. Validity time can coincide with transaction time if the fact is registered and deleted in the database exactly when it starts and ends in the real life but usually the two dimensions are not correlated. Validity time period is defined by the user regardless of when the fact is stored in the database.

In order to highlight the specific aspects of interest for Web sites, we concentrate on valid time which would suffice to show the main ideas. Transaction time will only partially be treated, as we store the timestamp of the modifications since we think it is a relevant information for the navigation of temporal data, however we do not completely cover the management of this temporal dimension. Rather than describing all technical details, our purpose is to define a methodology and a set of models for the development of data intensive web sites, with the ability of specifying temporal requirements at design time.

In a Web site, the motivation for valid time is similar to the one in temporal databases: we are often interested in describing not only snapshots of the world, but also histories about its facts. However, there is a difference: in temporal databases the interest is in storing histories and in being able to answer queries about both snapshots and histories, whereas in Web sites the challenge is on how histories are offered to site visitors, who browse and do not query. Therefore, this is a design issue, to be dealt with by referring to the requirements we have for the site. Also, since in many cases redundancy is even desirable in Web sites, we could have even both snapshot and histories.

Temporal Web sites require the management of temporal data and especially its collection. In this respect, it is worth noting that most Web sites are supported by applications that handle their data (and the updates to them). These applications are often implemented with the use of a Content Management System (CMS). We claim that the extension of CMSs with the explicit management (acquisition and maintenance) of time-related data can provide a significant contribution to our goal. This would obviously require the representation, in the CMS repository, of the temporal aspects of the information to be published.

This paper is aimed at giving a contribution to the claim that the management of time in Web sites can be effectively supported by leveraging on the experiences made in the database field, and precisely by the combination of the three areas we have briefly mentioned: temporal databases, content management systems and model-based development of Web sites. In particular, attention is devoted to models and design: (i) models in order to have a means to describe temporal features and (ii) design methods to support the developer in deciding on which are the temporal features of interest to the Web site user. The approach relies on a CASE tool that handles the various representations and transformations and so it gives significant support to the designer in the development of the various components. Many actual CMSs offer some temporal features (i.e. version control on pages, notifications on new content added) but the designer can only choose on whether to activate these features, that however do not allow an effective management of the time-varying elements of the site. We believe that time is a coordinate that should be managed at design time rather than by fixed features. What we are proposing here is a structured approach to the temporal management of a Web site that involves the specification of temporal aspects (with fine granularity) during the design process. Temporal requirements can be specified at various levels of abstraction, coherently with the design phase of the process (conceptual of data, conceptual of the Web site, logical of the Web site). We face the temporal management of site elements from different points of view. The designer can specify the elements for which the history of changes is of interest, how to treat versions and how to present them to the final user. The content administrator can exploit a temporal CMS, that we automatically generate through the proposed approach, with a set of features for a flexible management of the time coordinate in a Web site. The final user can navigate histories of data through the Web site.

The paper significantly extends the experiences in the Araneus project [1, 4, 8, 9] where models, methods and a CMS prototype for the development of data-intensive Web sites were developed. Indeed, we propose a logical model for temporal Web sites, a design methodology for them and a tool to support the process. A preliminary version of the tool has been demonstrated (see Atzeni and Del Nostro [10]), we refer here to a new version that also provides a CMS-style interface to support the user's modifications to the site content and gets temporal information from these actions.

The rest of the paper is organized as follows. Section 2 is devoted to a brief review of the aspects of the Araneus approach that are needed as a background. Then, Section 3 illustrates the temporal extensions for the models we use in our process and Section 4 the specific CMS features. In Section 5 the methodology with the associated tool is illustrated by means of an example of usage. Finally, in Section 6 we briefly sketch possible future developments. In the paper we use the notation where the term model has to be read as data model and the term schema has to be read as model.

2   The Araneus Models and Methodology

The Araneus approach (Merialdo et al. [4]) focuses on data-intensive Web sites and proposes a design process (with an associated tool) that leads to a completely automatic generation of the site extracting data from a database. The design process is composed of several steps each of which identifies a specific aspect in the design of a Web site. Models are used to represent the intensional features of the sites from various points of view:

1. the Entity Relationship (ER) model is used to describe the data of interest at the conceptual level (then, a translation to a logical model can be performed in a standard way, and is indeed handled in a transparent way by the associated tool);

2. a "navigational" variant of the ER model (initially called NCM and then N-ER in this paper) is used to describe a conceptual scheme for the site. The main constructs in this model are the major nodes, called *macroentities*, representing significant units of information, which consolidate concepts from the ER model (one or more entities/relationships), and navigation paths between macroentities, expressed as *directed relationships* (possibly with multiple paths and cycles). Nodes of an additional type, called *aggregations*, are used to define access structures for the hypertext. An example of aggregation is the home page of the site that aggregates links to access the content of the site;

3. a logical scheme for the site is defined using the Araneus Data Model (ADM), in terms of *page schemes*, which represent common features of pages of the same "type" with possibly nested attributes, whose values can come from usual domains (text, numbers, images) or be links to other pages.

The design methodology (sketched in Figure 1, see Atzeni et al. [8]), supported by a tool called Homer (Merialdo et al. [4]), starts with conceptual data design, which results in the definition of an ER scheme, and then proceeds with the specification of the navigation features, macroentities and directed relationships (that is, a N-ER scheme). The third step is the description of the actual structure of pages (and links) in terms of our logical model, ADM.

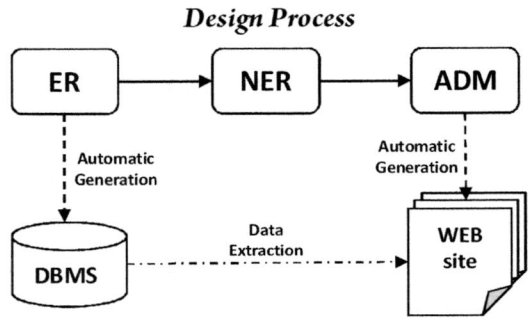

Fig. 1. The Araneus design process

Three simple schemes for the Web site of a University department, to be used in the sequel for comments, are shown in Figures 2, 3, and 4, respectively.

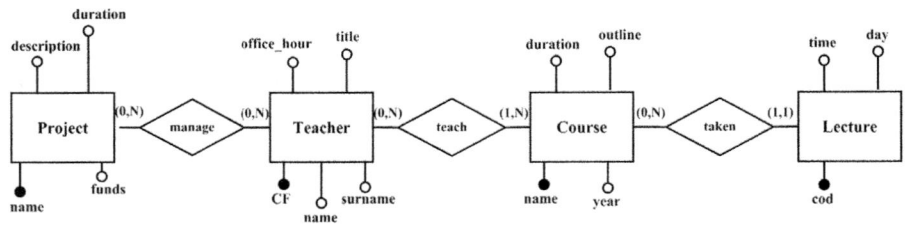

Fig. 2. ER schema for university Web site example

A fourth step is the specification of the presentation aspects, which are not relevant here. In the end, since all the descriptions are handled by the tool and the various steps from one model to the other can be seen as algebraic

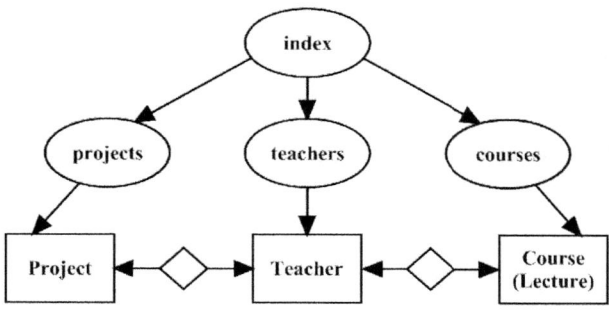

Fig. 3. N-ER schema for university Web site example

indexPage

projects ≡
toprojectsPage ◇

courses ≡
tocoursesPage ◇

teachers ≡
toteachersPage ◇

ProjectsListPage

ProjectList

name ≡
toProjectPage ◇

TeachersListPage

TeacherList

surname ≡
toTeacherPage ◇

CoursesListPage

CourseList

name ≡
toCoursePage ◇

ProjectPage

name ≡
description ≡
duration ≡
funds ≡
start ≡
end ≡

TeacherList

surname ≡
toTeacherPage ◇

TeacherPage

CF ≡
name ≡
surname ≡
office_hour ≡
title ≡

CourseList

name ≡
toCoursePage ◇

ProjectList

name ≡
toProjectPage ◇

CoursePage

name ≡
year ≡
duration ≡
outline ≡

TeacherList

surname ≡
toTeacherPage ◇

LectureList

time ≡
day ≡

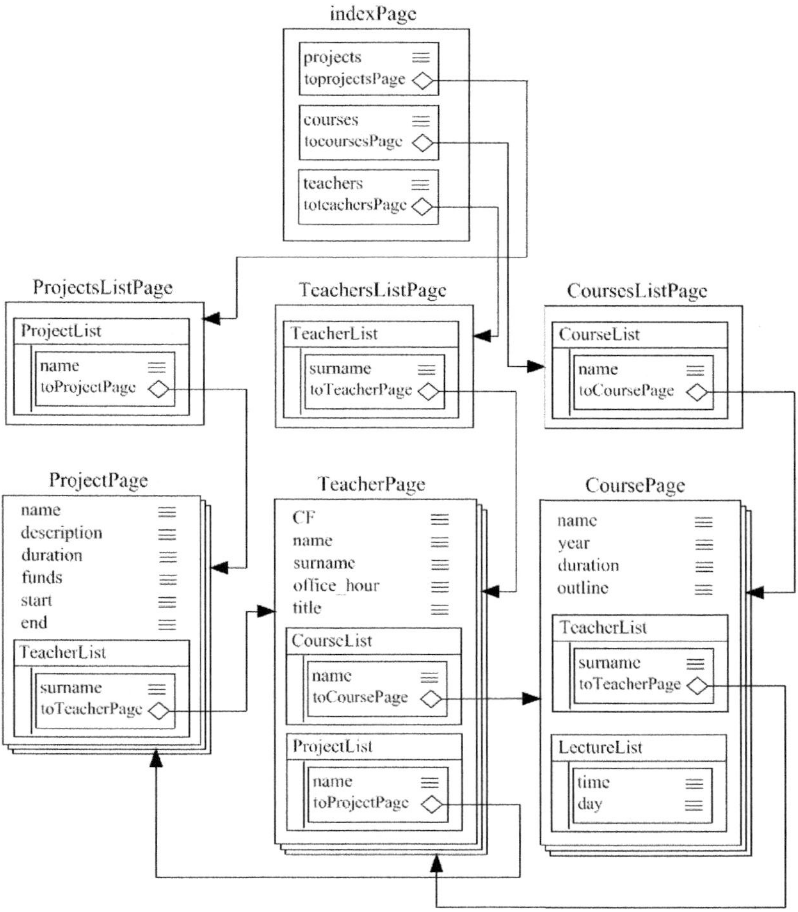

Fig. 4. ADM schema for university Web site example

transformations, the tool is able to generate, in an automatic way, the actual code for pages, for example in JSP or in plain HTML, with access to a relational database built in a natural way from the ER scheme.

3   Models for the Management of Temporal Aspects of Web Sites

Temporal aspects appear in all phases of the design process and therefore each of our models needs a suitable extension. We consider the points of view of the three types of users that are involved in a Web site content evolution: final users, who access the site and need tools to explore the content (and its changes), content administrators, who exploit the CMS and apply content modifications, and designers, who design the site. In our approach the designer can specify how temporality should be managed and presented to the final user and, at the end of the process, a temporal web site is automatically produced together with the related content management application that can be used by content administrators. Our goal is to have a "standard" Web site with features that simplify the management of temporal aspects. Considering that the most common use of the Web site will involve current values, the introduction of temporal aspects should not negatively impact the site structure complexity. We start with brief comments on the models we use for describing our data and then illustrate the conceptual and logical hypertext models that have been properly extended to allow the representation of temporal aspects.

3.1   Models for the Representation of Data

Different kinds of representations have been proposed to manage temporal aspects at a conceptual level, each of which with its specific features (see Gregersen and Jensen [11] for a survey). Some of the models represent a temporal object in the schema and allow designers to define temporal elements by a rather complex visual notation that needs a steep learning curve. Other approaches are based only on textual notation that is used to specify temporal properties beside the snapshot conceptual modeling. While the schema readability is not compromised, designers need to jump between the two models to know which concept has temporal features.

As illustrated by Artale et al. [15], there is a gap between the modeling solutions provided by researchers and the real needs of the designers. A conceptual model should be simple and expressive, in order to allow the representation of concepts, properties and relations by a visual interface with intuitive constructs. The diffusion of the ER model is based right on this characteristics. Therefore, in the conceptual design phase, we choose an extension of the ER model (hereinafter called T-ER) where temporal features are added to the scheme, by indicating which are the entities, relationships and attributes for which the temporal evolution is of interest. With our conceptual model proposal we don't aim at covering all the aspects that may arise when facing with temporal evolution of content but providing designers with some intuitive tools that allow the management of temporal aspects within a site design process. For each object in a scheme, the model allows to specify whether it is temporal. To keep the conceptual model as simple as possible we separate the specification of temporal elements from the definition of temporal properties for which we provide a textual notation. We briefly illustrate the main temporal constructs we have considered in our framework. A temporal object 0^T can be a single attribute, an entity, or a relationship. In the schema, temporal objects are identified by an uppercase T as a superscript. To allow designers to define a time granularity for an object, we provide the chronos construct with the syntax CHRONOS(G, 0^T) where G is the time granularity and 0^T is the temporal object. Examples of time granularity G are "day", "week", "month", "year". For example, if an object is defined as temporal with the month chronos, then one value per month will be considered for publishing. So looking at the so defined temporal schema, it can be immediately noticed which are the objects for which the evolution is of interest, keeping the schema simplicity without compromising readability. A separated textual notation is devoted to the refining of temporal properties, as is in the standard ER model where concepts are illustrated graphically while constraints are specified in a textual way. In Figure 5 an example usage of the notation is sketched.

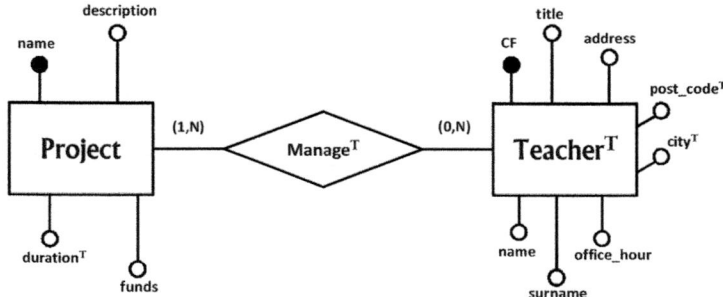

Fig. 5. Temporal notation example

Let's start a quick explanation of what the schema expresses. The designer wants to keep the *Project.duration* evolution, which can be changed, for example due to an extension of the ending time. Notice that in this model it is possible to specify temporal elements at different levels of granularity either for single attributes or for an entity (or a relationship) as a whole. Moreover, as illustrated in Figure 5, an entity and its attributes can be independently defined as temporal: this is the case for Teacher and its attributes. Temporality of the entity means that we are interested in its history: in the example, in the instants when an individual became a teacher in our school and when he quits or retires (but we will keep information after that). If an entity is not temporal, we are not interested in the history of the concept it represents. For the attributes, the interest in the evolution of values (the address, in the example).

If a single attribute is marked as temporal, its evolution is maintained with independently of other attributes. For example if every year the project manager can be changed then the designer specify that the CHRONOS for the Manage relation is Y (year). So in this case the notation is CHRONOS(Y, Manage^{T}). Since attributes do not exist without an entity, deleting an entity with temporal attributes results in loosing the attributes history.

The database used to handle the data for our Web sites is relational, as in the Araneus approach, with temporal features added to it (if using our tool the temporal features are generated automatically so the developer do not need any specific competence). Optimization aspects for temporal databases are beyond our focus, as they have been deeply studied [6], thus a simple relational schema for temporal tables has been chosen. When an entity is specified as temporal, then two timestamps are added to the relational representation, in order to define the validity time and a field to store the time when a modification is applied. The original primary key is therefore extended including the validity timestamps. For each temporal attribute an additional table is created to manage its modifications separately.

3.2   Models for the Representation of Web Sites

We use two models to describe the structure of a Web site at different levels of abstraction: the N-ER (data) model is used to express how concepts from the ER model are organized in a navigational scheme whereas the ADM (data) model is used to specify the actual structure of pages. The same distinction applies to their temporal extensions and (as they are tightly related) to the specific CMS aspects that concern modifications management. Therefore the conceptual representation of the Web site (N-ER model) is extended to allow the selection of the versions of interest for a concept. Then, in the logical model (ADM), new constructs are introduced to give the designer the possibility to choose how to organize versions in pages. Let us consider the two aspects in turn.

Temporal Aspects of Web Sites at a Conceptual Level. The temporal N-ER model allows the specification of whether temporality have to be managed for the concepts (macroentities and directed relationships) of interest for the site, and how. Details of versions management (eg. version integration) are out of the scope of this paper.

There are two main versioning aspects involved in this model. The first refers to the possibility that an object can be modified and the second concerns the inclusion of the evolution of changes in the site. These two points of view do not necessarily coincide. Based on the choices expressed in the T-ER model, different possibilities exist. If an object is identified as snapshot, it is here possible to choose whether it may change or not. Specifying here that a snapshot object is modifiable means that updates are allowed but the temporal database will not store the history of changes. In our example we could have that the *description* attribute of the *Project* entity is defined as snapshot in the T-ER model, because there is no interest in keeping track of its versions, but we want the system to allow changes and the Web user to be informed when the last changes happened (logical level aspects will be detailed in the next section). This is an example of whether a snapshot object should be considered as *modifiable* in the T-NER model.

In this model it is possible to define the temporality features for each of the temporal concepts (macroentities, direct relationships, and attributes). A concept can be defined as temporal if its origin in the T-ER scheme is a temporal component, but not necessarily vice versa. Therefore, we could have macroentities that are not defined as temporal even if they involve temporal elements, for example because the temporality is not relevant within the macroentity itself (indeed, Web sites often have redundancy, so an attribute or an entity of the ER scheme could contribute to various macroentities, and, even if temporal, it need not be temporal in all those macroentities). Consider the case where the designer needs to keep the history of all the modifications to the project information then the *Project* entity is defined as temporal in the T-ER model. The project data are used in two macroentities, *ProjectsListPage* and *ProjectPage*. The *ProjectsListPage* macroentity is just a list of projects with the project name as a link label, and therefore will not have any version management. The *ProjectPage* macroentity gives the user all the details about a project and it is here interesting to have a management of versions in order to let the user know about the changes.

It is worth noting that the approach is flexible and the solutions for the management of time that we propose here can be extended if others are needed.

For macroentities and attributes, a major facet is relevant here: which versions are of interest from the Web site conceptual point of view? We consider this as a choice from a set of alternatives, which currently include (i) none; (ii) the last version; (iii) all versions.

Till now we have considered temporal aspects of each concept separately (how to manage versions for each temporal element) but this need not be the only approach. An alternative would be navigation with respect to a specific instant, neglecting the others. Let us explain the idea by an example. Consider the following situation of the university Web site where both the Teacher page and the Course page are temporal: the teacher Smith page has two versions each of which with its validity interval as in Figure 6, the Database course given by Smith has the two versions illustrated in Figure 7 (dashed gray boxes highlight the changes).

If visiting the site we are on the Smith page we can select to view only information valid in a specific day. Then, we might want to keep on navigating with reference to that same day. Following the link to the Database course page, only data valid in the previously selected instant should be published. Navigating the site at the current time (without any temporal specification) the navigation path is shown in the upper part of Figure 8. Choosing the validity instant 04/04/2010 the correct navigation path is sketched in the lower part of Figure 8. This kind of navigation is called "time-based selective navigation" between pages.

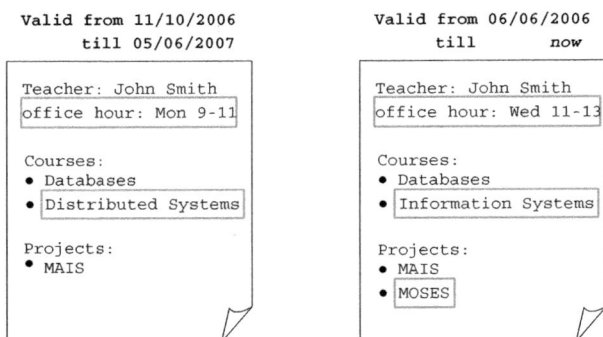

Fig. 6. Versions for the TeacherPage

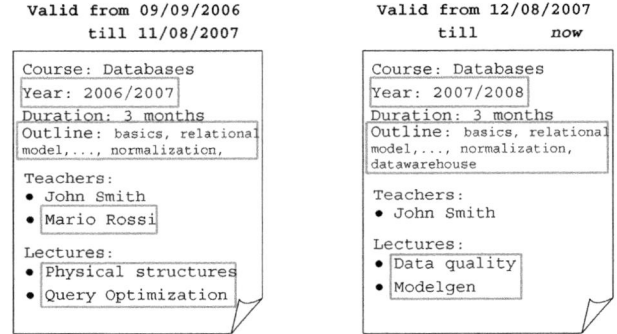

Fig. 7. Versions for the CoursePage

As another example of temporal navigation, let us consider a university Web site where people have the personal page with the publications section. It could be interesting to trace the career level of the author at the time a certain article has been written. If the Web site has not been designed considering the temporal coordinate it would not be possible to answer to this kind of request.

Temporal Aspects of Web Sites at a Logical Level. The logical design of a temporal Web site has the goal of refining the description specified by a temporal N-ER scheme, by introducing all the details needed at the page level: how concepts are organized in pages and how the history of temporal elements is actually published. The temporal extension of ADM (hereinafter *T-ADM*) includes all the features of ADM (and so allows for the specification of the actual organization of attributes in pages and the links between them), and those of the higher level models (the possibility of distinguishing between temporal and non-temporal page schemes, and for each page, the distinction between temporal and non-temporal attributes; since the model is nested, this distinction is allowed at various levels in nesting, apart from some technical limitations), and some additional details, on which we concentrate. A major choice here is the

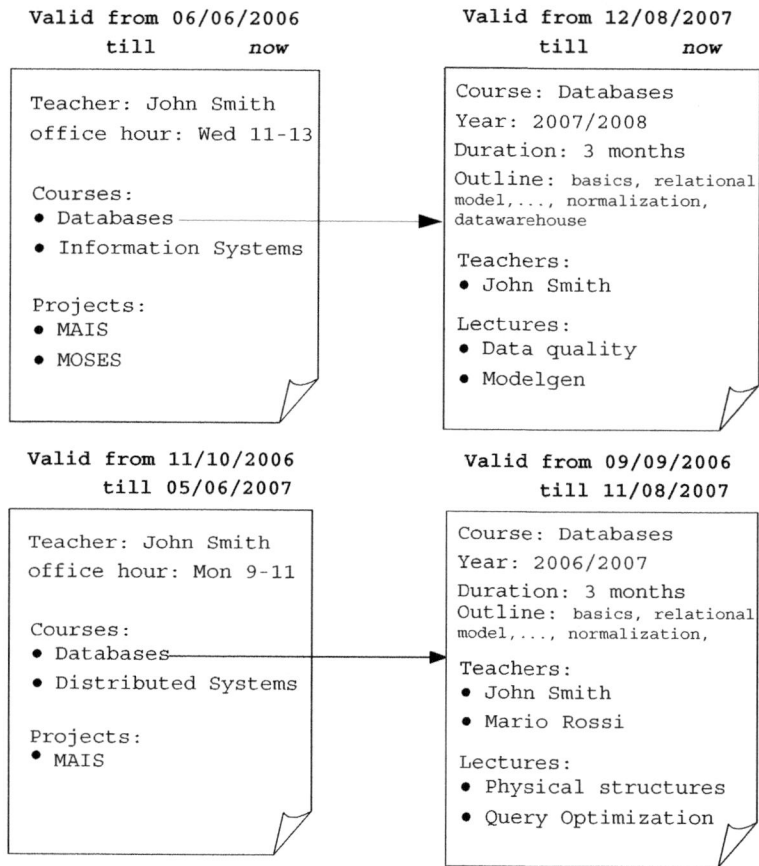

Fig. 8. Two examples of temporal navigation

implementation of the temporal requirements specified at the conceptual level. A version is generated when a temporal element, as defined in the T-ER model (attribute, entity, relationship) is modified. Basing on the choices expressed in the temporal N-ER model, in T-ADM we can define the details of the versions presentation.

As this model can be obtained as a translation from the higher level models, the temporal choices expressed on the previous models drive the various alternatives offered:

(a) No versioning
(b) Last version
(c) All versions in the same page
(d) All versions in separate pages
(e) Time based selective navigation

In case (a) no version management is required; this is actually a consequence of the decision of not considering versions in the temporal N-ER model. When only the last version is of interest, then the designer expresses choice (b). This can regard either temporal elements, when the designer does not want to publish previous versions but only the current value, or snapshot elements in the case the modification information are needed in the page. When the N-ER design choice for a temporal element is to manage all versions, the alternative (c) allows the possibility to include them together in the same T-ADM page scheme. This means that the designer gives all the versions the same importance thus the user will catch them in a single view. If the current and the previous versions have different browsing priority, it is possible to separate the "current value" from the previous ones, correlated by means of links, choosing alternative (d). Various browsing structures, which will be detailed later on, allow different ways to publish versions according to the designer choices. A completely different organization is the "time-based selective navigation", represented by option (e). In this case the user selects, for a page, the instant of interest and sees the corresponding valid versions. The navigation between temporal pages can then proceed keeping reference to that instant.

Additional features allow for the highlight of recent changes (on a page or on pages reachable via a link). The above features are expressed in T-ADM by means of a set of constructs, which we now briefly illustrate.

We first illustrate the T-ADM extensions that allow the designer to choose which meta data have to be published with versions: CREATOR, MODIFIER, DESCRIPTION, as we briefly illustrate in the following:

CREATOR. Represents the creator of a content. This information can just be associated with the current value of an information or (introducing a bit of redundancy) with each of its versions.

MODIFIER. This attribute is related to a version and represents the author of the change to a data element.

DESCRIPTION. It can be related to an element that can be changed (temporal or snapshot with the differences illustrated in the previous section) to publish the reason of the modification.

We also have two constructs representing two different temporal pieces of information that can be associated with the content version:

LAST MODIFIED. This is a special, predefined attribute used to represent the date/time (at the granularity of interest) of the last change applied to a temporal element. This is a rather obvious, and widely used technique, but here we want to have it as a first class construct offered by the model (and managed automatically by the support tool) and also we think it should be left to the site designer to decide which are the pages and/or attributes it should be actually used for, in order to be properly informative but to avoid overloading.

VALIDITY INTERVAL. This is another standard attribute that can be associated with any temporal element.

The next is a major feature of the model, as it is the basis for the time-based selective navigation:

TIME POINT SELECTOR. It can be associated with pages and with links within them, in such a way that navigation can proceed with reference to the same time instant; essentially, in this way the user is offered the site with the information valid at the specified instant.

Another feature is used to highlight a link when the destination is a page that includes temporal information which has recently (according to a suitable metric: one day, one week, or whatever the designer chooses) changed:

TARGET CHANGED. This property can be used in association with LAST MODIFIED to add the time the modification has been applied.

The TARGET CHANGED feature is illustrated in Figure 9: a Department page (source) has a list of links to teacher pages (target). In a teacher page the office hours have been modified. When the user visits the department page he is informed which teacher pages have been modified (and when) so he can follow the link to check what is new. The example refers to just one source and one target page, but things may become more interesting when we consider non-trivial hypertextual structures: this gives the opportunity to propagate this kind of information through a path that leads to the modified data (see Figure 10). When a new lecture is introduced, then both the teacher and the department page are informed (and highlight the change) so the user can easily know which are the site portions with modified data.

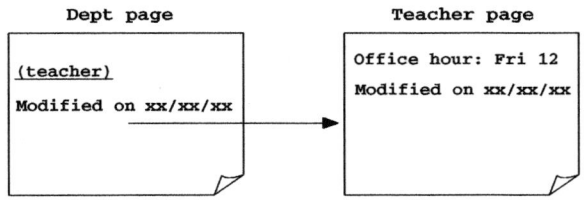

Fig. 9. The TARGET CHANGED feature

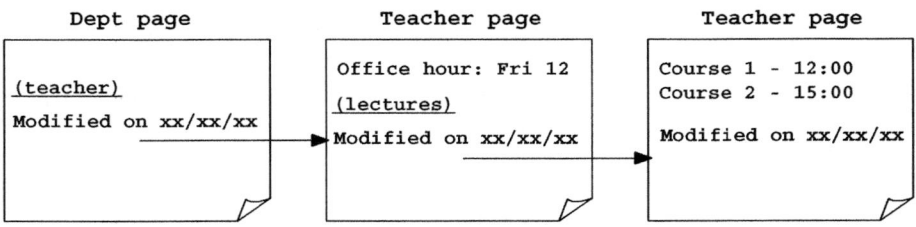

Fig. 10. The TARGET CHANGED feature along a path

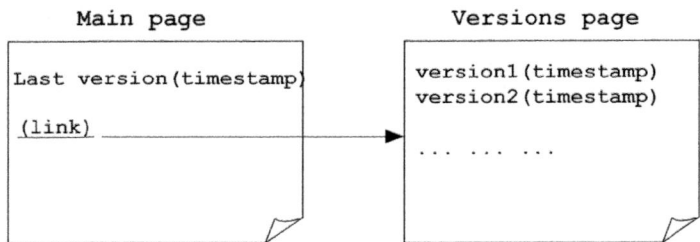

Fig. 11. The SIMPLE VERSION STRUCTURE pattern

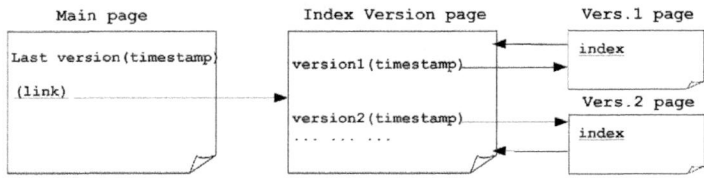

Fig. 12. The LIST VERSION STRUCTURE pattern

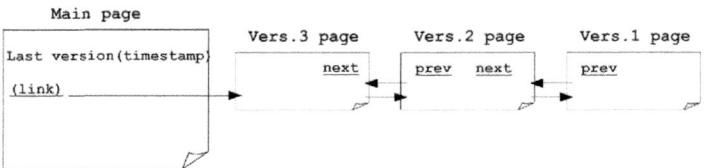

Fig. 13. The CHAIN VERSION STRUCTURE pattern

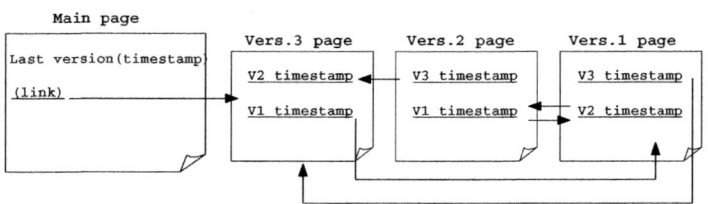

Fig. 14. The SUMMARY VERSION STRUCTURE pattern

We now illustrate the additions brought to the model to implement the different ways versions can be presented to the user:

REVISION LIST. This feature allows for the specification that all versions of a temporal element are shown in the same page as a list of revisions.

LINK TO VERSIONS. This is a special type of link that has as a target a VERSION STRUCTURE (to be illustrated shortly), handling the versions of a temporal element. It is used when the designer chooses to have just the last version in the main page and the others held in other pages.

VERSION STRUCTURE. These are "patterns" for pages and page schemes, used to organize the different versions of a temporal element and referred to by the LINK TO VERSIONS attribute. There are various forms for this construct involving one or more pages:

- SIMPLE VERSION STRUCTURE: a single page presenting all the versions for the temporal element with timestamps.
- LIST VERSION STRUCTURE: an "index" page with a list of links labeled with the validity intervals that point to pages showing the particular versions and include links back to the index.
- CHAIN VERSION STRUCTURE: this is a list of pages each of which refers to a specific version. It is possible to scan versions in chronological order, by means of the "previous" and "next" links available in each page.
- SUMMARY VERSION STRUCTURE: similar to the previous case but the navigation between versions is not chronological. Each version page has a list of links that works as an index to all versions.

As we mentioned earlier, the designer is guided during the design process. The extension of the Araneus methodology, that involve the temporal management proposed in this paper, will be referred to as T-Araneus. The tree of available choices, along the T-Araneus process, that can be expressed at each abstraction level is sketched in Figure 15.

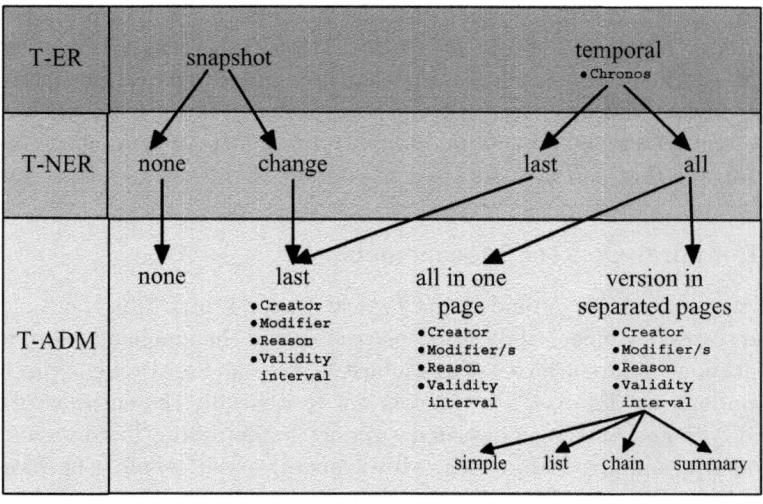

Fig. 15. The tree of available choices

4   CMS Support to T-Araneus

The need for Content Management Systems (CMSs) as fundamental tools for handling the information to be published in Web sites was recognized soon after the establishment of the Web as a strong communication medium [12–14]. In data-intensive Web sites this would mean that the updates to the database should be controlled in a rigorous way. From our point of view, it is worth noting some features usually available in CMSs (i.e. the management of the workflow, the management of user and their associated roles, and the assignment of responsibilities and permissions to different categories of users for different types of contents) and a special requirement often neglected by them.

One of the goals of a CMS is to support users in the changes to the content, allowing the management of the whole life cycle of these contents also through workflow management capabilities.

The most relevant feature of these systems, which is very important for our purposes, is that they can handle the history of a site and of its contents (and the associated responsibilities: "who changed a piece of content and when?"). However CMSs usually have a lack of flexibility in handling content granularity. In particular they consider pages or documents as the granularity of interest (e.g. Wikis may have little temporal features but only provide support for tracking different versions of pages). Instead, we believe that, in many cases, especially for data-intensive Web sites, different levels of granularity are needed: individual pieces of data, corresponding to atomic fields in databases, a table in the related database, the whole page or even the entire site.

Starting from the Araneus tool, whose approach is discussed in Section 2, used to generate a site from the definition of the aforementioned models, we have developed a new CASE tool that also handles the temporal information of a site. Moreover, to take profit from the CMS features this tool now generates both the site and the temporal CMS that is required to manage the site itself. In other words, an instance of our CMS is generated along with the Web site.

In particular we extend the functionality related to the management of temporal features that can be associated to content at different level of granularity.

4.1   Temporal Content Management

In our work, we have extended the typical features of CMSs focusing on two aspects: the management of data-intensive sites and the management of temporal information at different levels of granularity, from the finest (single piece of data in the underlying database) to the coarsest (potentially the entire site).

Our CMS guarantees the needed support for handling histories of data at the needed granularity. Moreover, with a fine granularity, also the structure of pages in the CMS could become complex and delicate to design, but it can be supported by the models and methodology we have discussed earlier.

The high level models we explained in the previous sections are fully supported through the CMS features. Given the fine granularity of data and the possible redundancy in the data in the site, non-trivial design choices arise for the definition of the pages of the CMS: which pages are needed (for example, the instructor for a course appears in various pages; do we want to have CMS pages for updating it corresponding to each of the Web site pages or just one of them suffices?), with what attributes, and so on. At the same time, by incorporating the CMS in our methodology, we can derive some decision on structures by considering the temporal properties of pieces of data: if an attribute cannot change, then there is no need for a form to handle the updates.

The use of our CMS can greatly improve the integration of time into the World Wide Web. This should enable the user to reconstruct historical content and follow its evolution.

From the CASE tool interface it is possible to design the entire site, therefore the designer can define all temporal features from the different scheme views (from ER to N-ER and finally to ADM). Therefore the site is automatically generated from the data stored in the underlying database and the necessary tables for storing the temporal features are also created. Then the CMS pages are generated and from this point all contents of the site can be managed through the CMS functionalities.

As the CMS is automatically generated on the basis of the designer's choices at the time he/she specifies the temporal features of the site, these choices also modify both the presentation forms (all versions in a page, versions in separated linked pages) and the update/deletion forms of the content.

An important observation is that, by managing all updates, our CMS can easily produce and maintain the meta-data of interest for the generation of the pages of our sites, some of which have been mentioned in the previous sections: the author and modifier of a piece of data, the time information for a given change, the motivation for it.

Moreover, our temporal CMS supports some typical and useful features of a standard CMS, namely: workflow management, user administration.

One of the most typical features of a CMS is the possibility to access the content at various levels and with different rights, through a user management system. Our system is based on the definition of different classes of users, which can be managed through the CMS interface. In particular there are three main classes: (i) Administrator: the users of this class can have access to all managing features. They can define new users or user classes and define the respective authorizations. (ii) Editor: these users will be the managers of individual sites, and can be considered as administrators of their sites; (iii) Author: they are users that can create or modify only the contents of a specific area of a site (assigned by an Editor). Each change creates a temporal version of a content according to the chosen aforementioned methods. Our CMS has also a Workflow System that enables the collaborative management of content and temporal information. An *activity* represents the needed operation to complete the workflow task; namely we distinguish between (i) *atomic activities* i.e. a set of operations required by the workflow, that can be combined so that they appear to the rest of the system to be a single operation, (ii) *complex activities* that imply the existence of nested workflows. Unlike traditional workflow systems that are available to a CMS, our workflow activities may affect the contents manipulation at different levels of granularity, from the coarsest (an entire Web page) to the finest (the single record in the database).

4.2   Architecture of the System

In Figure 16 a sketch of the architecture of our working prototype is presented.

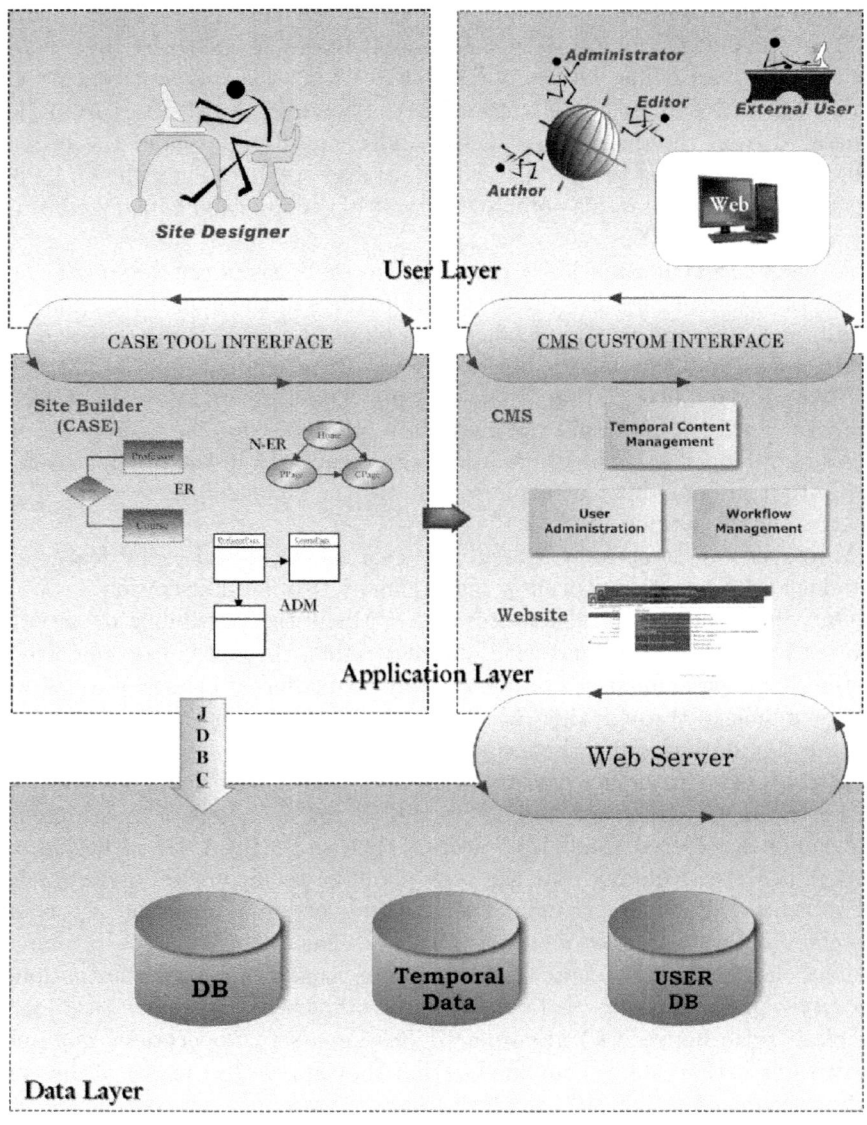

Fig. 16. Architecture of the system

The architecture is characterized by three different layers, namely: i) repository, ii) application and iii) user.

The repository is managed through a relational DBMS (MySQL in our prototype implementation but it can be easily replaced by any other relational databases) that stores content, temporal information and user information. The communication with the CASE tool is realized via JDBC, while the connection with the CMS (and obviously the underlying sites) is made through an Apache Web Server.

Therefore we have the Application layer that implements the business logic to manage contents and their temporal metadata. As we can see in the figure the elements this level includes the CASE tool through which the designer can define the structure of a site and the temporal choices for the management of the contents. On the basis of these choices the CASE tool generates the site along with the CMS for the management of the site itself.

Our tool can manage the whole contents life-cycle, by means of definition of the three main models (ER, N-ER and ADM) through the CASE tool, see Figure 17 for a screenshot of the site design interface. Our CMS is fully accessible through a standard Web browser and is completed with the traditional features of workflow and user management.

The upper layer is the user layer. The CASE tool can be accessed offline by the designer of the site with a J2SE interface. With the CMS a user access and manage the contents and temporal data through a standard Web browser. Each user, depending on the different role can access to the CMS functionality through a customized interface. We distinguish between internal (site designers, administrators, authors, etc.) and external (or final) users of the CMS.

5   An Example Application

Let us now exemplify the design process by referring to the example introduced in Section 2 which, despite being small, allows us to describe the main issues in the methodology. We also sketch how the tool we are implementing supports the process itself. Rather than showing a complete example, where it would be heavy to include all the temporal features, we refer to a non-temporal example, and comment on some of its temporal extensions.

The first step is to add temporal features to the snapshot ER schema. Let us assume that the requirements specify that we need: (i) to know the state (with all attribute values) of the entity *Project* when a change is applied to one or more attribute values; and (ii) to keep track of the modifications on the *office_hour* attribute for the *Teacher* entity. The first point means that the whole *Project* entity needs to be temporal, whereas for the second point, indeed, the designer has to set the temporality only for the attributes *office_hour* in the *Teacher* entity.

In Figure 18 we illustrate the portion of interest of the resulting T-ER schema: the elements tagged with T are those chosen as temporal.

With respect to the conceptual design of the navigation, we have already shown in Section 2 the overall N-ER scheme. Let us concentrate here on the

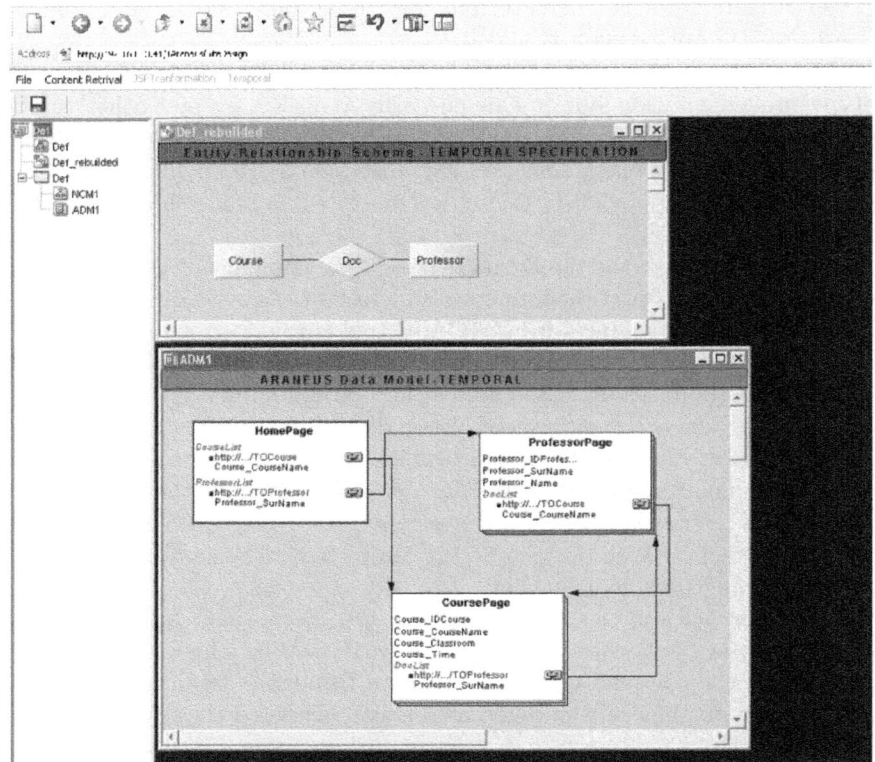

Fig. 17. A screenshot of the CASE tool design interface

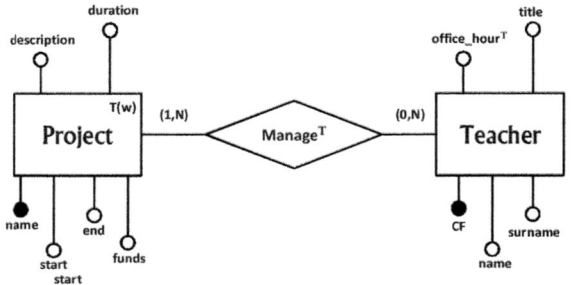

Fig. 18. The example T-ER schema

temporal features: at this point it is possible to choose how to manage versions for each macroentity and attribute. We have defined the *Project* entity as temporal in the T-ER model so the temporal database will handle the modifications but we don't want the site to show all versions so we choose here to have only the last version with a timestamp in the *Project* macroentity. It will be possible in the future to change this choice and add versions for projects by simply modifying this property. For the temporal attribute *office_hour* in *Teacher* we want all the versions to be managed and the snapshot attribute *title* should be modifiable via CMS. In Figure 19 the temporal N-ER scheme is shown with explicit indication of the version management choices (with the following codes: AV: all versions; LV: last version; SU: snapshot updatable).

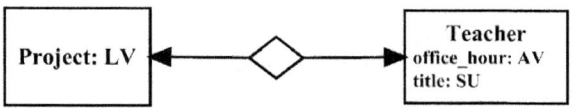

Fig. 19. Temporal features in the N-ER model

Let us then consider the logical design. As we have discussed in Section 2, a standard ADM scheme can be automatically generated as an algebraic transformation based on the conceptual models (it can then be restructured if needed). During this automatic generation, for each temporal element in a page scheme, it is possible to specify how to present versions starting from the choices made in the N-ER scheme. It is also possible, in this phase, to specify whether to include meta-information and how to show it in pages. On the basis of the requirements, we could decide to handle versions for *CoursePage* by means of a CHAIN VERSION STRUCTURE. For the *TeacherPage* page scheme we could choose to handle the last version in the main page for the *office_hour* attribute with an associated SIMPLE VERSION STRUCTURE page presenting all versions. The *Teacher* page should also present information about the last update for the *title* attribute.

The T-ADM page scheme for the *TeacherPage* is illustrated in Figure 20. The *office_hour* attribute is associated with the VALIDITY INTERVAL information and a LINK TO VERSION that points to the SIMPLE VERSION STRUCTURE page scheme presenting all the versions each with a VALIDITY INTERVAL. The snapshot attribute *title* has instead been associated with a LAST MODIFIED meta-information.

At the end of the design process, the tool can be used to generate the actual site, which can be static (that is, plain HTML) or dynamic (JSP); actually some of the features (such as the time point selector) are allowed only in the dynamic environment. It is worth noting that the sites we generate with our approach are completely standard, as they require common http servers, with just ordinary JSP support.

The temporal Web site generation comes with the automatic generation of the CMS to manage the contents. Initially a predefined user, with administrator privileges, is created to allow the designer to define all the users/groups and rights of access to information.

It can be observed, from the analysis of the aforementioned example, that the prosed temporal extensions could lead to higher operating costs. These extra costs can arise at design time. However, the development of our prototype has shown that the biggest costs are proportional to the amount of details we want

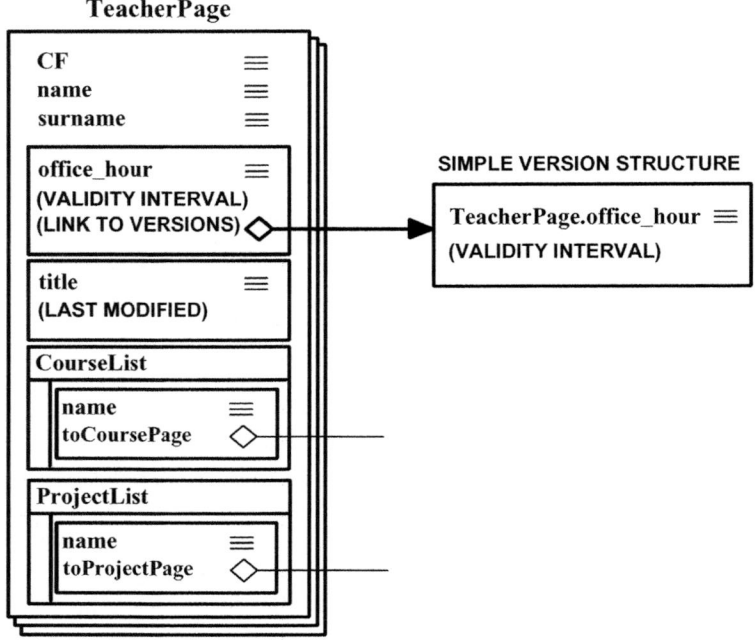

Fig. 20. A T-ADM page scheme

to address. Indeed, the higher the amount of contents we want to manage in a temporal way, the higher will be the costs. The increase of complexity only regards the structural level and slightly affects the overall performance.

6   Conclusion and Future Works

We noticed that web sites designers often ignore (or superficially treat) information versioning management with negative consequences on the user experience during web sites navigation. Users cannot access history of time-evolving data or even can't know when an information is still valid or is no more up-to-date. We believe that the temporal coordinate has to be treated as a design issue and the designer should choose whether to manage it and how, with a fine level of granularity on what is temporal. In this paper we have proposed a methodology for the development of data-intensive web sites, together with models, to help developers describing temporal features at design time. At the end of the design process a web site can be produced (together with the corresponding CMS) that facilitates site users and content administrators in navigating and managing time-varying information. We have focused on one of the aspects of interest for the management of temporal evolution: the content. Other dimensions are obviously of interest for real-world, complex Web sites, and we plan to consider them in the near future. They include: (i) the presentation; (ii) the hypertext structure; (iii) the database structure. Among them, the most challenging is probably the last one: as we consider data intensive Web sites, the hypertext structure is obviously strongly related to the database structure so it could be very important to keep track of the evolution of the schema. If you change the ER schema (and, as a consequence, the underlying database schema), for example deleting an entity and a relationship, it can result in a change in the hypertext structure and/or the presentation. Essentially, this would be a variation of a maintenance problem, with the need to keep track of versions.

References

1. Atzeni, P., Mecca, G., Merialdo, P.: To weave the Web. In: VLDB 1997, Proceedings of 23rd International Conference on Very Large Data Bases, Athens, Greece, August 25-29, pp. 206–215. Morgan Kauffman, Los Altos (1997)
2. Ceri, S., Fraternali, P., Bongio, A., Brambilla, M., Comai, S., Matera, M.: Designing Data-Intensive Web Applications. Morgan Kauffman, Los Altos (2002)
3. Fernandez, M., Florescu, D., Levy, A., Suciu, D.: Declarative specification of Web sites with Strudel. VLDB Journal 9, 38–55 (2000)
4. Merialdo, P., Atzeni, P., Mecca, G.: Design and development of data-intensive web sites: The Araneus approach. ACM Trans. Inter. Tech. 3, 49–92 (2003)
5. Jensen, C., Snodgrass, R.: Temporal data management. IEEE Transactions on Knowledge and Data Engineering 11, 36–44 (1999)
6. Jensen, C., Snodgrass, R.: Temporal Database Encyclopedia of Database Systems, pp. 2957–2960 (2009)
7. Snodgrass, R.: Developing Time-Oriented Database Applications in SQL. Morgan Kaufmann, Los Altos (1999)
8. Atzeni, P., Merialdo, P., Mecca, G.: Data-intensive web sites: Design and maintenance. World Wide Web 4, 21–47 (2001)
9. Merialdo, P., Atzeni, P., Magnante, M., Mecca, G., Pecorone, M.: Homer: a model-based case tool for data-intensive web sites. In: Proceedings of the 2000 ACM SIGMOD International Conference on Management of Data, Dallas, Texas, USA, May 16-18, p. 586, ACM (2000)
10. Atzeni, P., Del Nostro, P.: T-Araneus: Management of Temporal Data-Intensive Web Sites. In: Hwang, J., Christodoulakis, S., Plexousakis, D., Christophides, V., Koubarakis, M., Böhm, K. (eds.) EDBT 2004. LNCS, vol. 2992, pp. 862–864. Springer, Heidelberg (2004)
11. Gregersen, H., Jensen, C.: Temporal entity-relationship models—a survey. IEEE Transactions on Knowledge and Data Engineering 11, 464–497 (1999)
12. Boiko, B.: Content Management Bible. Wiley Publishing Inc. (2002)
13. Addey, D., Ellis, J., Suh, P., Thiemecke, D.: Content Management Systems. Glasshaus (2002)
14. Nakano, R.: Web Content Management - A Collaborative Approach. Addison-Wesley (2001)
15. Artale, A., Parent, C., Spaccapietra, S.: Evolving objects in temporal information systems. Annals of Mathematics and Artificial Intelligence, vol. (50), pp. 5–38. Kluwer Academic Publishers (2007)

16. Brambilla, M., Ceri, S., Comai, S., Fraternali, P.: A CASE tool for modelling and automatically generating web service-enabled applications. Int. J. Web Eng. Technol. 2(4), 354–372 (2006)

17. Ceri, S., Brambilla, M., Fraternali, P.: The History of WebML Lessons Learned from 10 Years of Model-Driven Development of Web Applications. In: Borgida, A.T., et al. (eds.) Conceptual Modeling: Foundations and Applications. LNCS, vol. 5600, pp. 273–292. Springer, Heidelberg (2009)

18. Böhlen, M.H., Gamper, J., Jensen, C.S.: Multi-Dimensional Aggregation for Temporal Data. In: Ioannidis, Y., Scholl, M.H., Schmidt, J.W., Matthes, F., Hatzopoulos, M., Böhm, K., Kemper, A., Grust, T., Böhm, C. (eds.) EDBT 2006. LNCS, vol. 3896, pp. 257–275. Springer, Heidelberg (2006)

19. Raisinghani, M.S., Klassen, C.: Temporal Databases. In: Encyclopedia of Database Technologies and Applications, pp. 677–682 (2005)

Homogeneous and Heterogeneous Distributed Classification for Pocket Data Mining

Frederic Stahl[1], Mohamed Medhat Gaber[1], Paul Aldridge[1], David May[1],
Han Liu[1], Max Bramer[1], and Philip S. Yu[2]

[1] School of Computing, University of Portsmouth
Portsmouth, PO1 3HE, UK
[2] Department of Computer Science, University of Illinois at Chicago
851 South Morgan Street, Chicago, IL 60607-7053, USA

Abstract. Pocket Data Mining (*PDM*) describes the full process of analysing data streams in mobile ad hoc distributed environments. Advances in mobile devices like smart phones and tablet computers have made it possible for a wide range of applications to run in such an environment. In this paper, we propose the adoption of data stream classification techniques for *PDM*. Evident by a thorough experimental study, it has been proved that running heterogeneous/different, or homogeneous/similar data stream classification techniques over vertically partitioned data (data partitioned according to the feature space) results in comparable performance to batch and centralised learning techniques.

1 Introduction

Thanks to continuing advances on mobile computing technology, more and more data mining applications are running on mobile devices such as 'Tablet PCs', smart phones and Personal Digital Assistants (PDAs). The ability to make phone calls and send SMS messages nowadays seems to be merely an additional feature rather than the core functionality of a smart phone. Smart phones offer a wide variety of sensors such as cameras and gyroscope as well as network technologies such as Bluetooth, and Wi-Fi with which a variety of different data can be generated, received and recorded. Furthermore smart phones are computationally able to perform data analysis tasks on these received, or sensed data such as data mining. Many data mining technologies for smart phones are tailored for data streams due to the fact that sensed data is usually received and generated in real time and due to the fact that limited storage capacity on mobile devices requires that the data is analysed and mined on the fly while it is being generated or received. For example the Open Mobile Miner (OMM) tool [25] allows the implementation of data mining algorithms for data streams that can be run on smart phones.

Existing data mining systems for smart phones such as MobiMine [19] or VEDAS [20] and its commercial version *MineFleet* [21,23] are some examples of systems that deploy data stream mining technology to mobile phones. However to our knowledge all existing data mining systems for mobile devices either

A. Hameurlain et al. (Eds.): TLDKS V, LNCS 7100, pp. 183–205, 2012.

facilitate data mining on a single node or follow a centralised approach where data mining results are communicated back to a server which makes decisions based on the submitted results. Constraints that require the distribution of data mining tasks among several smart phones are, large and fast data streams, subscription fees to data streams, and data transmission costs in terms of battery and bandwidth consumption. The data transmission cost can be lowered by processing parts of the same data stream locally on different smart phone devices that only collaborate only by exchanging local statistics, or locally generated data mining models rather than raw data. The collaborative data mining on smart phones and 'Tablet PCs' facilitated by building an ad hoc network of mobile phones will allow to build significantly useful analysis tasks, however this area remains widely unexplored.

In this paper we describe and evaluate the Pocket Data Mining (PDM) framework, coined and proven to be computationally feasible in [3]. PDM has been built as a first attempt to explore collaborative and distributed data mining using stream mining technologies [6], mobile software agents technologies [7,8] and embedded programming for mobile devices such as smart phones and 'Tablet PCs'. The main motivation in developing PDM is to facilitate the seamless collaboration between users of mobile phones which may have different data, sensors and data mining technology available. The usage of mobile agent technology is motivated by the agent's autonomous decentralised behaviour, which enables PDM to be applied on highly dynamic problems and environments with a changing number of mobile nodes. A second motivation for using mobile agent technology is the communication efficiency of mobile agent based distributed data mining [9,10]. A general agent based collaborative scenario could look like the following. A mobile device that has a data mining task sends out a mobile agent that roams the network of mobile devices and collects for the data mining task useful information, such as which mobile devices have which data sources and/or sensors available and which data mining technologies are embedded on these devices. Next a further agent is sent out to consult the data mining task relevant mobile devices (in terms of their data sources, sensors and data mining technology) and uses the collective information to synthesize the data mining task.

Possible applications for PDM comprise:

– Stock market analysis tasks for investors and brokers, brokers can retrieve real time stock market data anytime anywhere they want using smart phones and can perform data mining on this data in order to support decisions to sell or buy shares [11]. However brokers may only want to subscribe to data of companies they are directly interested in, as the data transfer is expensive in terms of bandwidth consumption, processing power and battery life. Hence locally installed data mining technology may not pick up and learn direct/indirect dependencies between the subscribed shares and non subscribed shares. Collaborative data mining using PDM can overcome these limitations by sharing local models rather than data.

- Recent budget cuts of the British coalition government due to the current economic crisis could lead to a reduction of about 60000 police officers [26]. Reduction in police staff will have to be compensated. PDM could help streamlining the process of knowledge acquisition on the crime scene. The crime scene investigators could form an ad hoc network using their smart mobile phones. They could capture pictures, video data and finger prints as well as any other sensory data on the crime scene or from online data sources. If the task is to know more information about an aspect of the crime, the distribution of tasks could be that one device is to do an Internet search, another is to take pictures and a further one may retrieve data from other sensors close to the crime scene for example CCTV cameras. Without PDM the information recorded would have to be formatted and entered into a central data server and a data analyst would have to be employed to evaluate the gathered information, whereas with PDM the information could be fused together in real-time to give insights and knowledge about the crime.
- Sensors of smart phones can collect in situ continuously data about the healthy condition of a patient, for example some earphones of mobile phones can read a person's blood pressure [27], the accelerometer could detect physical activity, also the body temperature could be recorded and the fact that the mobile phone is used indicates that the user is conscious and probably well. Behaviour patterns can be mined from this sensory data. Nurses and 'mobile medical staff' could be equipped with mobile devices as well, if the nurse is idle, then the mobile phone can send out a mobile agent that roams the network of 'patients' and makes a decision where the nurse is needed most and instructs the nurse to go there. This decision may take the health status, the location of the patient and the nurse into account as well as the nurse's particular area of expertise.

The growing demand for commercial knowledge discovery and data mining techniques has led to an increasing demand of classification techniques that generate rules in order to predict the classification of previously unseen data. Hence classification rule induction is also a strong candidate technology to be integrated into the PDM framework. For example the classification of data streams about the stock market could help brokers to make decisions whether they should buy or sell a certain share. Also in the area of health care, classification of streaming information may be beneficial, for example the smart phone may record various health indicators such as the blood pressure and/or the level of physical activity and derive rules that may indicate if a patient needs medical attention, the urgency and what kind of medical attention is needed. For this reason a version of PDM that incorporates two strong data stream classification technologies has been created and is evaluated in this paper.

The paper is organised as follows. Section 2 highlights some related work in the area of distributed data mining on mobile devices; Section 3 describes the PDM framework in the context of classification rule induction on data streams; Section 4 evaluates several configurations of PDM comprising different classifiers; Section 5 highlights ongoing work on PDM and future developments; some concluding remarks can be found in Section 6.

2   Related Work

The topic of this research paper lies in the intersection of distributed data mining, mining data streams and mobile software agents.

Distributed data mining has been surveyed thoroughly by Park and Kargupta in [22]. Mainly, there are two broad categories of distributed data mining, namely, homogeneous and heterogeneous. It has to be noted that this categorisation is made in reference to the attributes. Homogeneous distributed data mining refers to the process of mining the same set of attributes over all the participating nodes. On the other hand, heterogeneous distributed data mining refers to mining different sets of attributes in each participating node. In this paper, our focus is on the vertically partitioned data, i.e., the heterogeneous distributed data mining scenario. This choice has been made to provide a more realistic scenario to the applications discussed in this paper. It is worth noting that in this paper, we use the terms homogeneous and heterogeneous to refer to the data mining algorithms used in the process, where homogeneous refers to the use of only one data mining algorithm across all the participating nodes, and heterogeneous refers to the use of different data mining algorithms.

Mining data streams, on the other hand, is a more recent topic of research. A concise review of the area by Gaber et al is given in [6]. A more detailed review is given in [28]. The area is concerned with analysis of data generated in a high speed relative to the state-of-the-art computational power, with a constraint of real-time demand of the results. Hundreds of algorithms have been proposed in the literature addressing the research challenges of data stream mining. Notable success of the use of Hoeffding bound to approximate the data mining models for streaming data has been recognised [13]. The two-stage process of online summarisation and offline mining of streaming data, proposed by Aggarwal el al [1,2], has been also recognised as a feasible approach to tackle the high data rate problem. Addressing both the resource constraints and high speed aspects of data stream mining has been addressed by Gaber et al [31,29,30] by proposing the algorithm granularity approach. The approach is generic and could be plugged into any stream mining algorithm to provide resource-awareness and adaptivity. Mining data stream algorithms is at the heart of our *pocket data mining* framework.

Finally, mobile software agents are computer programs that autonomously and intelligently move from one node to the other to accomplish its task. Potential applications and obstacles of this technology have been detailed in [7,8]. The use of mobile agent technology for distributed data mining has been recognised, as an alternative paradigm to the client/server technologies for large databases. A cost model has been developed by Krishnaswamy et al [12], suggesting a hybrid approach to distributed data mining, combining both client/server and mobile agent paradigms. Our choice of mobile agent paradigm in this research project has been due to the fact that our approach follows a peer-to-peer computation mode, and also that centralisation of the stream data mining in the mobile computing environment is infeasible.

3   PDM: Pocket Data Mining

PDM describes the collaborative data mining of data streams using mobile devices and mobile agent technology which is executed on an ad hoc mobile network. Section 3.1 illustrates the basic framework and workflow of PDM whereas Section 3.2 highlights a particular implementation of PDM using two different classifiers.

3.1   Architecture

The basic architecture is depicted in Figure 1. The mobile device that has a data mining task and utilises PDM to solve it is called the task initiator. PDM consists of three generic software agents that may be mobile and thus able to move between mobile phones within the ad hoc mobile network [3].:

- (Mobile) **A**gent **M**iners (AM) are distributed over the ad hoc network. They may be static agents used and owned by the user of the mobile devices. Or they may be mobile agents remotely deployed by the user of a different mobile device. They implement the basic stream mining algorithms; however they could also implement batch learning algorithms if required by the application.
- Mobile **A**gent **R**esource **D**iscoverers (MRD) are mobile agents that are used to roam the network in order to discover for the data mining task relevant data sources, sensors, AMs and mobile devices that fulfil the computational requirements. They can be used to derive a schedule for the Mobile Agent Decision Makers described below.
- Mobile **A**gent **D**ecision **M**akers (MADM) can move the mobile devices that run AMs and consult the AMs in order to retrieve information or partial results for the data mining task. The MADMs can use the schedule derived by the MRDs.

Algorithm 1 describes the basic data mining workflow that PDM employs collaboratively. The task initiator forms an ad hoc network of participating mobile devices within reach. Next the task initiator starts the MRD agent which will roam the network searching for data sources that are relevant for the data mining task, and for mobile devices that fulfill the computational requirements (battery life, memory capacity, processing power, etc.). While the MRD is collecting this information it will decide on the best combination of techniques to perform the data mining task. On its return to the task initiator the MRD will decide which AMs can and need to be deployed to remote mobile devices. There

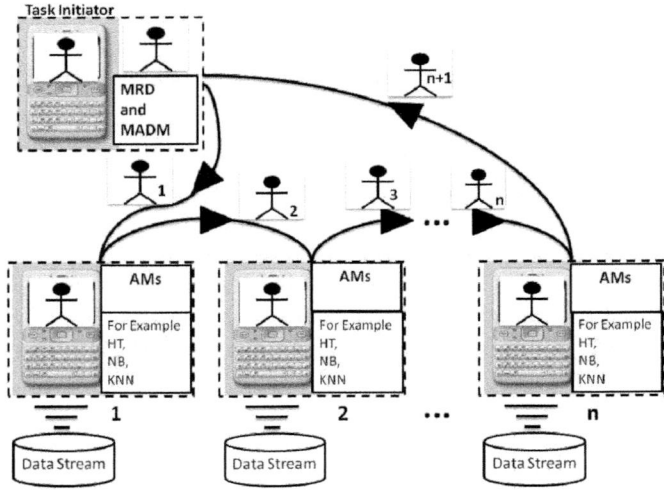

Fig. 1. The PDM Architecture

might be restrictions, some mobile phone owners may not allow alien AMs to be deployed, for example they may have limited computational capacity such as battery life, their data might be confidential etc. Concerning the confidentiality issues the owner of the mobile device may still allow its own AMs to be consulted as he will have control over its own AMs and thus about which information they release to alien MRD and MADM agents. The AMs are executed concurrently as indicated by the parallel for loop (parFor) in Algorithm 1. Finally the task initiator starts the MADM agent with the schedule provided by the MRD agent. The MADM agent visits the in the schedule listed AMs and will use their model in order to gain information for their decision making process. Finally on the return to the task initiator the MADM agent will make a collective decision based in the information gathered from the AMs distributed in the network.

Algorithm 1. PDM's collaborative data mining workflow

Task Initiator: Form an ad hoc network of mobile phones;

Task Initiator: start *MRD* agent;

MRD: Discover data sources, computational resources and techniques;

MRD: Decide on the best combination of techniques to perform the task;

MRD: Decide on the choice of stationary AMs and deploy mobile *AMs*;

Task Initiator: Start *MADM* agent with schedule provided by the *MRD*;

parFor $i = 1$ **to** $i = number\ of\ AMs$ **do**

AM_i: starting mining streaming data until the model is used by the MADM agent.

end parFor

The current implementation of PDM offers AMs that implement classifiers for data streams, in particular the Hoeffding trees and the Naive Bayes classifiers which will be described in Section 3.2. It is assumed that the AMs are subscribed to the same data stream, however potentially to different parts of it. For example in the context of the stock market application outlined in Section 1, the mobile device may only have data about shares available in which the user of the mobile device is interested in, or more general the mobile device may only be subscribed to specific features of the data stream. A broker may train its own AM classifier on the data he has subscribed to, this could be for example by updating a model which is based on classes 'buy', 'sell', 'do not sell' and 'undecided' whenever he makes a new transaction. He may also use the current model to support his decisions to 'buy' or 'sell' a share. However, if the broker is now interested in buying a new share he has not much experience with, thus he may be interested in what decisions other brokers are likely to make in the same situation. Other brokers may not want to disclose their actual transactions but may share their local AM or even allow alien AMs to be deployed and for this the brokers can use PDM. With the current version of PDM the data mining workflow outlined in Algorithm 1 may look like the following, where the mobile device of the broker interested in investing in a new share is the task initiator. In the steps below and elsewhere in the paper, if we refer to a PDM agent hopping, we mean that the agent stops is execution, is transferred by PDM to a different mobile device and resumes its execution on this device. Also in the steps below it is assumed that the ad hoc network is already established:

1. Task Initiator: Send a MRD agent in order to discover mobile devices of brokers that have subscribed to relevant stock market data, i.e. data about the shares the broker is interested in.
2. The MRD agent hops from mobile device to mobile device and if it finds a device subscribed to relevant data it memorises the device and also if there are any useful AMs already available. If there are no useful agents it will memorise if the device allows alien agents to be deployed.
3. The MRD agent returns to the task initiator. From there the MRD agent will remotely deploy relevant AMs to mobile devices in its list that have relevant data but no relevant AMs however, allow alien AMs to be deployed remotely.
4. Once all AMs are deployed the MRD agent composes a schedule of all relevant classifier AMs subscribed to relevant data and passes it on to the MADM agent.
5. The MADM agent loads the data about the new shares the broker is interested in and starts hopping to each AM in the schedule.
6. On each AM the MADM agent hands over the 'shares data' to the AM and asks to classify it for example with class 'buy', 'do not buy', 'sell', 'do not sell' or 'undecided. The MADM may also retrieve some estimate how reliable the AMs thinks its classification is, for example its local classification accuracy.
7. Once the MADM returns to the task initiator it may employ a majority voting on the collected classifications from each AM or a weighted majority

---

Algorithm 2 Hoeffding tree induction algorithm.

1: Let HT be a tree with a single leaf (the root)
2: **for all** training examples **do**
3:      Sort example into leaf l using HT
4:      Update sufficient statistics in l
5:      Increment n_l, the number of examples seen at l
6:      **if** $n_l \bmod n_{min} = 0$ **and** examples seen at l not all of same class **then**
7:          Compute $\overline{G}_l(X_i)$ for each attribute
8:          Let X_a be attribute with highest $\overline{G}_l$
9:          Let X_b be attribute with second-highest $\overline{G}_l$
10:         Compute Hoeffding bound $\epsilon = \sqrt{\frac{R^2 \ln(1/\delta)}{2n_l}}$
11:         **if** $X_a \neq X_\emptyset$ **and** $(\overline{G}_l(X_a) - \overline{G}_l(X_b) > \epsilon$ **or** $\epsilon < \tau)$ **then**
12:             Replace l with an internal node that splits on X_a
13:             **for all** branches of the split **do**
14:                 Add a new leaf with initialized sufficient statistics
15:             **end for**
16:         **end if**
17:     **end if**
18: **end for**

---

voting incorporating the AMs local accuracy (we will call this the AMs weight). The outcome of the (weighted) majority voting is used as recommendation for the broker to the investment in the new share.

3.2   Implementation of PDM Using Distributed Hoeffding Trees and Distributed Naive Bayes for Mining Data Streams on Mobile Devices

PDM in its current version offers two AMs for classification tasks on data streams. One of the AMs implements the Hoeffding Tree classifier [13] and one that implements the Naive Bayes classifier. The AM that employs the Hoeffding Tree classifier uses the Hoeffding Tree implementation from the **M**assive **O**nline **A**nalysis (MOA) tool [4] as outlined by Bifet and Kirkby [5] and shown in Algorithm 2. Hoeffding tree classifiers have been designed for high speed data streams.

The Naive Bayes classifier has been originally developed for batch learning; however its incremental nature makes it also applicable on data streams. Again the AM employing the Naive Bayes classifier uses the Naive Bayes implementation from the MOA tool [4]. Naive Bayes is based on the Bayes Theorem [14] which states that if C is an event of interest and $P(C)$ is the probability that event C occurs, and $P(C|X)$ is the conditional probability that event C occurs under the premise that X occurs then:

$$P(C|X) = \frac{P(X|C)P(C)}{P(X)}$$

The Naive Bayes algorithm uses the Bayes Theorem to assign to a data instance to the class it belongs to with the highest probability.

4 Evaluation of PDM

This paper examines the PDM's applicability of classification rule induction to data streams. Three different configurations of PDM have been thoroughly tested. One PDM configuration solely based on Hoeffding Tree AMs, the second configuration of PDM is solely based on Naive Bayes AMs and the third configuration is a mixtures of both Hoeffding Tree and Naive Bayes AMs. Section 4.1 outlines the general experimental setup, Section 4.2 outlines the experimental results obtained using only Hoeffding Tree AMs, Section 4.3 outlines the experimental results obtained using only Naive Bayes AMs and Section 4.4 outlines the experimental results obtained using a mix of Hoeffding tree and Naive Bayes AMs.

4.1 Experimental Setup

The two classifier AMs use the Hoeffding Tree and Naive Bayes implementations from the MOA toolkit [4] with the reasoning that MOA is based on the WEKA [15] data mining workbench and thus supports the usage of the well known .arff data format. PDM is also built on the well known **J**ava **A**gent **D**evelopment **E**nvironment (JADE) [16]. JADE agents are hosted and executed in *JADE containers* that can be run on the mobile devices and PCs. JADE agents can move between different JADE containers and thus between different mobile devices and PCs. As JADE agents can be developed on PCs and run on both PCs and mobile phones it is possible to develop and evaluate PDM on a LAN of PCs. The used LAN consists of 9 workstations with different software and hardware specifications and is connected with a CISCO Systems switch of the catalyst 2950 series. In the configurations of PDM examined in this paper 8 machines were either running one Hoeffding Tree or one Naive Bayes AM. The 9th machine was used as the task initiator, however any of the 8 machines with AMs could have been used as task initiator as well. The task initiator starts the MADM in order to collect classification results from the AMs.

The data streams for PDM have been simulated using the datasets described in Table 1. The datasets have been labelled with test 1 to 6 for simplicity when referring experiments to a particular data stream. The data for test 1, 2, 3 and 4 have been retrieved from the UCI data repository [17] and datasets 5 and 6 have been taken from the Infobiotics benchmark data repository [18]. All datasets are stored in the .arff format and the data stream is simulated by taking a random data instance from the .arff file and feeding it to the AM. Instances may be selected more than once for training purposes.

Table 1. Evaluation Datasets

Test Number	Dataset	Number of Attributes	Number of Instances
1	kn-vs-kr	36	1988
2	spambase	57	1999
3	waveform-500	40	1998
4	mushroom	22	1978
5	infobiotics 1	20	≈ 200000
6	infobiotics 2	30	≈ 200000

As mentioned in Section 3 each AM may be subscribed to only a subset of the total feature space of a data stream, we call this a vertically partitioned data stream. For example a stock marked broker may only subscribe to data about companies he is interested in investing, or a police officer may only access data he has clearance for. Even if a user of a mobile device may have access to the full data the owner of the device may not want or be able to subscribe to 'for him' unnecessary features for computational reasons, such as bandwidth consumption, also the more data is processed by AMs the more power they will consume, or simply the processing time of the data stream is longer the more features are streamed in and need to be processed and higher subscription fees may be imposed. Yet the current subscription may be insufficient for classifying new data instances. However the task initiator can send a MADM with the unclassified data instances. This MADM visits and consults all relevant AMs that belong to different owners that may have subscribed to different features that are possibly be more relevant for the classification task.

The MADM collects predictions from each AM for each unclassified data instance and the estimated 'weight' (accuracy) of the AM, which it uses to decide on the final classification. In the *PDM* framework each AM treats a streamed labelled instance either as train or as test instance with a certain probability which is set by the owner of the AM. The default probability used in the current setup is 20% for the selection as a test and 80% for the selection as a training instance. Each training instance is put back into the stream and may be selected again as training instance, this allows to simulate endless data streams with reoccurring patterns. The test instances are used to calculate the 'weight' of the AM. The AM also takes concept drifts into account when it calculates its 'weight' by defining a maximum number of test instances to be used. For example if the number of test instances it 20 and there are already 20 test instances selected then the AM replaces the oldest test instance by the newly incoming test instance and recalculates the 'weight' using the 20 test instances.

After the MADM finished consulting all AMs in its schedule it returns to the task initiator and uses the local predictions from each AM and the AMs weights in order to derive a final classification using a 'weighted majority voting'. For example for the classification of one data instance, if there are three AMs, *AM1*, *AM2* and *AM3*. *AM1* predicts class A and has a weight of 0.57, *AM2* also predicts class *A* and has a weight of 0.2 and *AM3* predicts class *B* and has a weight of 0.85. The MADM's 'weighted' prediction for class A is $0.57A + 0.2A = 0.77A$ and for class B $0.85B = 0.85B$. Thus the MADM yielded the highest weighted vote for classification B and will label the concerning instance with class B.

The user of PDM can specify which features its AM shall subscribe to, however in reality we may not know the particular subscription, thus in the experimental set-up each AM subscribes to a random subset of the feature space. In particular experiments with each AM holding 20%, 30% and 40% of the total feature space have been conducted.

The terminology that is used in Sections 4.2, 4.3 and 4.4 is explained below:

- The **weight** refers to the local accuracy of the AM calculated using randomly drawn test instances from the local data stream.
- **MADM's accuracy** or **PDM's accuracy** is the accuracy achieved by the MADM using the test dataset classified by 'weighted majority voting' by the MADM.
- **local accuracy** is not to be confused with the weight. The local accuracy is the actual accuracy that a particular AM achieved on classifying the MADM's test data. This accuracy is only calculated for evaluation purposes, it would not be calculated in the real application as the real classifications of the MADM's test set would be unknown.
- the **average local accuracy** is calculated by averaging the local accuracies of all AMs. The average accuracy is used to show if the 'weighted majority voting' performs better than simply taking a majority vote.

4.2   Case Study of PDM Using Hoeffding Trees

The datasets listed in Table 1 are batch files. Using batch files allows us to induce classifiers using batch learning algorithms and thus to compare PDM's classification accuracy to the ideal case of executing batch learning algorithms on the whole datasets using all attributes. In particular the C4.5 [24] and Naive Bayes batch learning algorithms have been used from the WEKA workbench [15]. The choice of C4.5 is based on its wide acceptance and use; and to the fact that the Hoeffding tree algorithm is based on C4.5. The choice of Naive Bayes is based on the fact that it is naturally incremental, computationally efficient and also widely accepted.

In general it is expected that the more features the AMs have available the more likely it is that they achieve a high classification accuracy and thus the more likely it is that the MADM achieves an high classification accuracy as well. Yet some features may be highly predictive and others may not be predictive and even introduce noise. Thus in some cases having more features available may decrease the AMs and thus PDM's accuracy. 70% of the data instances from each dataset in Table 1 have been used to simulate the local data stream and the remaining 30% have been used as test data in order to evaluate PDM's and respectively MADM's accuracy. All experiments outlined in this paper have been

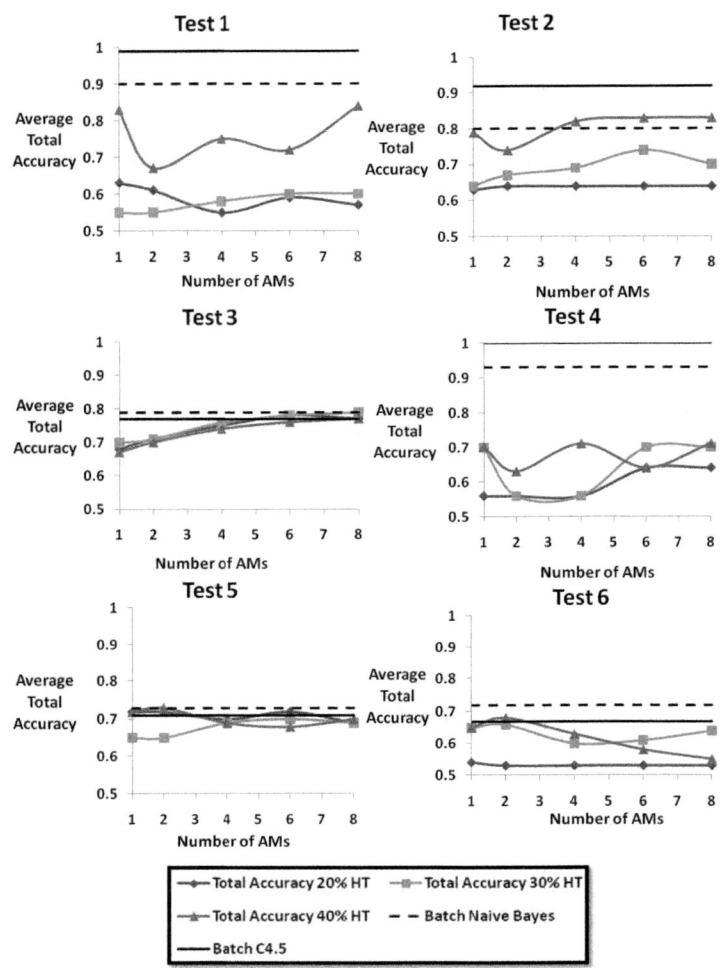

Fig. 2. PDM's average classification Accuracy based on Hoeffding Trees

conducted 5 times, the average local accuracy of each AM has been calculated and recorded as well as PDM's or respectively MADM's average accuracy.

Figure 2 shows PDM's average classification accuracy plotted versus the number of AMs visited. The experiments have been conducted for configurations where all AMs either subscribe to 20%, 30% or 40% of the features of the data stream. The features each AM subscribes to are selected randomly thus some AMs may have subsets of their features in common and some not. That two or more AMs have features in common is realistic, for example for the stock market broker application briefly outlined in Section 1. Two brokers may be interested in the 'Compaq' share but only one of them may be interested in the 'Hewlett-Packard' share and one in the 'Microsoft' share.

The largest difference between Naive Bayes's accuracy and C4.5 is for test 2 where Naive Bayes's accuracy is 80% and C4.5 91%, otherwise both batch learning algorithms achieve similar accuracies. Concerning PDM's accuracy based on Hoeffding trees it can be seen that PDM generally achieves accuracies above 50% for all datasets. In general PDM configurations with AMs using just 20% of the feature space generally perform much worse than configurations with 30% or 40% which can be explained by the fact that predictive features are more likely not to be selected. In some cases, for example for test 2 it seems that configurations of PDM with 30% achieve a higher accuracy than configurations with 40% which can be due to the fact that with subscribing to 40% of the features it is also more likely that non predictive features that also introduce noise are selected compared with subscribing to 30%. In general it can be observed that if subscribing to 30% instances achieves better results than subscribing to 40% instances the difference in accuracy between both configurations is not very large. In general PDM achieves accuracies close to the batch learning algorithms C4.5 and Naive Bayes, notably in tests 3 and 5 but also for the remaining tests PDM achieves close accuracies to those of Naive Bayes and C4.5. In general PDM based on Hoeffding trees achieves acceptable classification accuracy in most cases.

Varying the number of AMs generally is dependent on the dataset used. Highly correlated attributes in one dataset would only need small number of AMs and vice versa.

Figure 3 compares PDM's accuracy (achieved by the MADM through 'weighted majority voting') with the average local accuracy of all AMs versus the number of AMs visited. Each row of graphs corresponds to one of the tests in Table 1 and each column of graphs corresponds to a percentage of features the AMs are subscribed to. The lighter line in the graphs is the accuracy of PDM and the darker line is the average local accuracy of all AMs. PDM's accuracy is in most cases higher or even better than the average local accuracy, hence the MADM's 'weighted majority voting' achieves a better result compared with simply taking the average of the predictions from all AMs.

4.3   Case Study of PDM Using Naive Bayes

A further configuration of PDM solely based on Naive Bayes AMs has been evaluated the same way as PDM solely based on Hoeffding trees has been. PDM solely based on Naive Bayes is expected to produce similar results compared with PDM solely based on Hoeffding trees evaluated in Section 4.2.

Figure 4 presents the data obtained of PDM solely based on Naive Bayes the same way as Figure 2 does for PDM solely based on Hoeffding trees. Again the experiments have been conducted for configurations where all AMs either subscribe to 20%, 30% or 40% of the features of the data stream. The features each AM subscribes to are selected randomly thus some AMs may have subsets of their features in common and some not. Concerning PDM's accuracy based on Hoeffding trees it can be seen that PDM generally achieves accuracies above 50% for all datasets. Similar compared with Figure 2 PDM configurations with

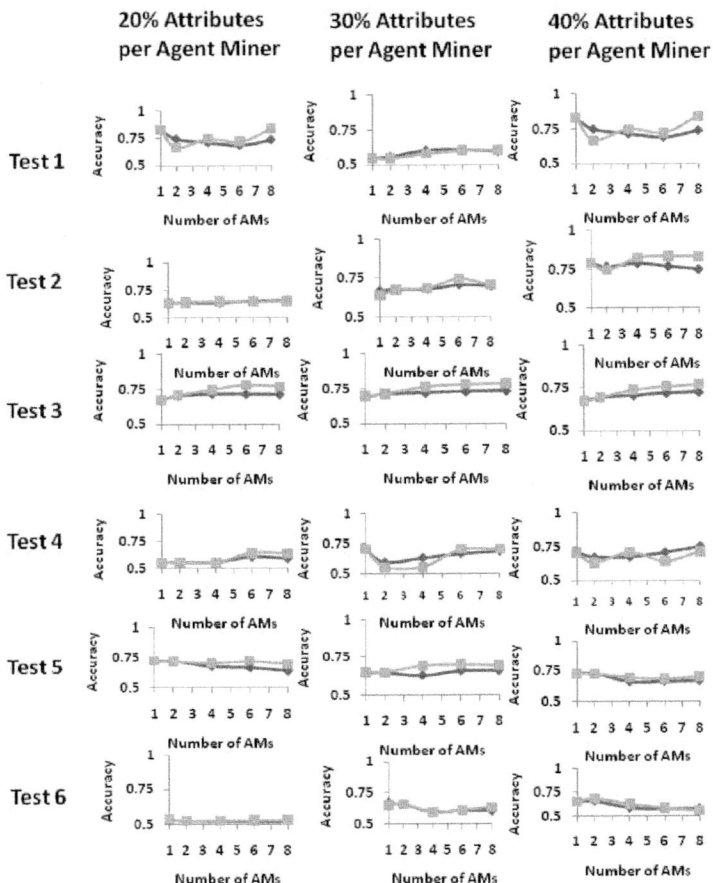

Fig. 3. PDM's average classification accuraciy versus the average local accuracy of the AMs with Hoeffding Trees

AMs using just 20% of the feature space generally perform much worse than configurations with 30% or 40% which can be explained by the fact that predictive features are more likely not to be selected. Yet in some cases, for example for test 2 it seems that configurations of PDM with 30% achieve a higher accuracy than configurations with 40% which can be due to the fact that with subscribing to 40% of the features it is also more likely that non predictive features that introduce noise are selected compared with subscribing to 30%. In general PDM achieves accuracies close to the batch learning algorithms C4.5 and Naive Bayes, notably in tests 3, 4, 5 and 6. However also for the remaining tests PDM achieves close accuracies to those of Naive Bayes and C4.5. In general PDM based on Hoeffding trees achieves acceptable classification accuracy in most cases.

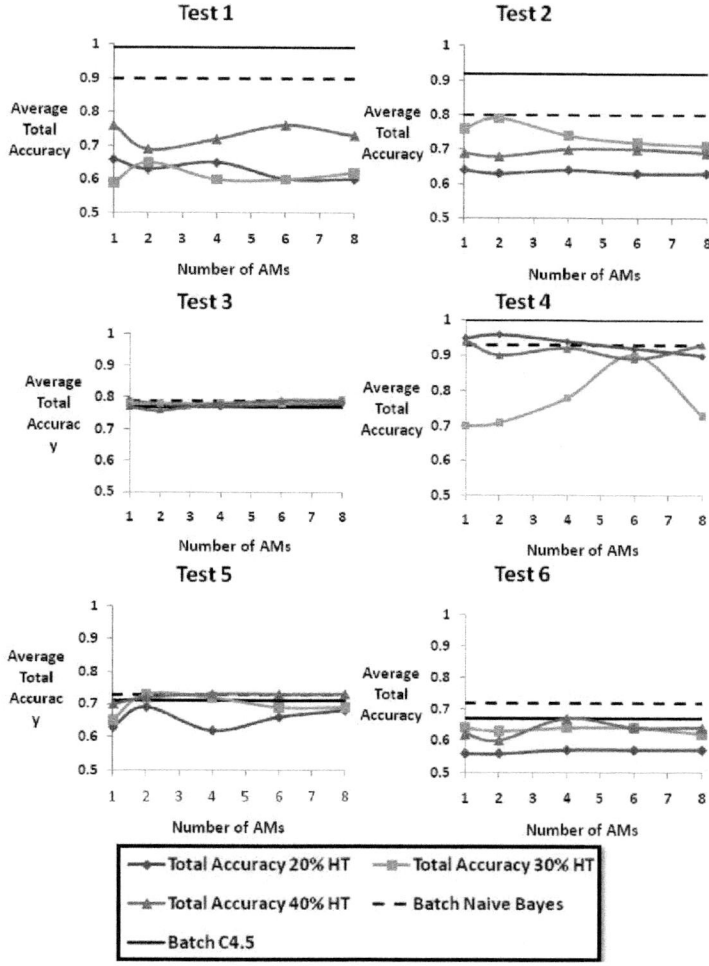

Fig. 4. PDM's average classification Accuracy based on Naive Bayes

Similar to the previous set of experiments, varying the number of AMs generally is dependent on the dataset used. Highly correlated attributes in one dataset would only need small number of AMs and vice versa.

Figure 5 analogous to Figure 3 opposes PDM's accuracy (achieved by the MADM through 'weighted majority voting') and the average local accuracy of all AMs versus the number of AMs visited. Each row of graphs corresponds to one of the tests in Table 1 and each column of graphs corresponds to a percentage of features the AMs are subscribed to. The lighter line in the graphs is the accuracy of PDM and the darker line is the average local accuracy of all AMs. Similar to the Hoeffding tree results PDM's accuracy is in most cases higher or even better

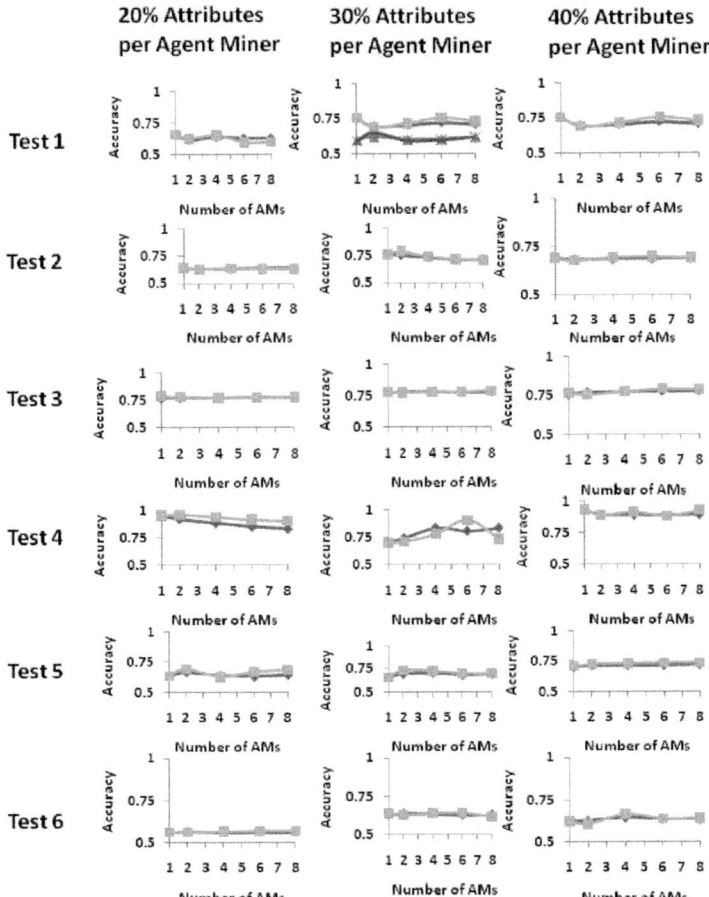

Fig. 5. PDM's average classification accuracy versus the average local accuracy of the AMs with Naive Bayes

than the average local accuracy, hence the MADM's 'weighted majority voting' either achieves a better result than simply taking the average of the predictions from all AMs.

4.4   Case Study of PDM Using a Mix of Hoeffding Trees and Naive Bayes

Figure 6 highlights the accuracies of the two PDM configurations solely based on Hoeffding and solely based in Naive Bayes for different numbers of visited AMs. The bars in the figure are in the following order from left to right: Total accuracy of PDM with Hoeffding Trees with 20% attributes; total accuracy of *PDM* with Hoeffding Trees with 30% attributes; total accuracy of PDM with Hoeffding Trees with 40% attributes; total accuracy of PDM with Naive Bayes with

20% attributes; total accuracy of PDM with Naive Bayes with 30% attributes; total accuracy of PDM with Naive Bayes with 40% attributes; accuracy for batch learning of Naive Bayes with all attributes; and finally accuracy for batch learning of C4.5 with all attributes.

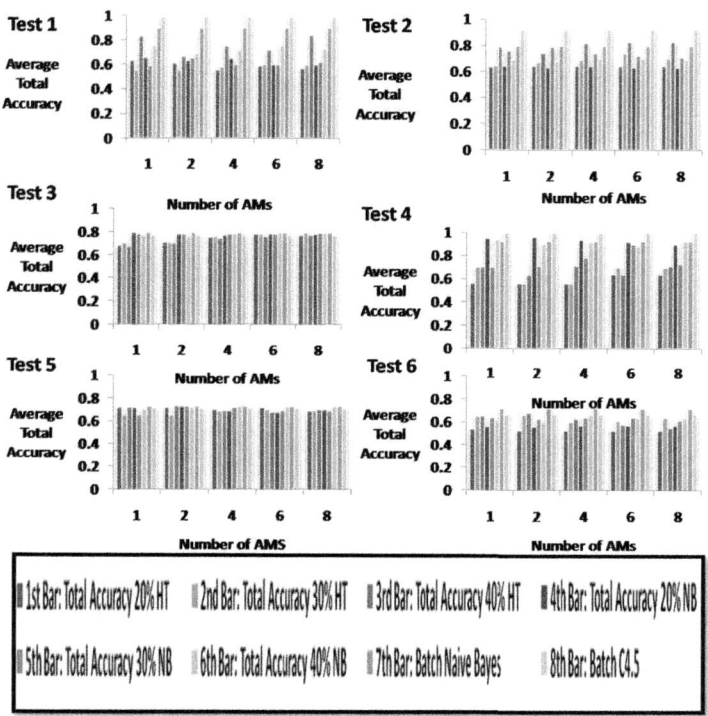

Fig. 6. PDM's average classification accuracy for both configurations with Hoeffding Trees and Naive Bayes

On tests 3 and 5 both configurations of PDM achieve an almost equal classification accuracy. Also for tests 1, 2, 4 and 6, the classification accuracies of PDM are very close for both configurations and there does not seem to be a bias towards one of the classifiers used. Hence a heterogeneous configurations of PDM with a mixture of both classifiers would be expected to achieve a similar performance than a homogeneous configuration solely based on Hoeffding Trees or Naive Bayes. Such a heterogeneous set-up of AM classifiers would also be a more realistic set-up as owners of mobile devices may use their individual classification techniques tailored for the data they subscribed to.

In order to show that a heterogeneous set-up of AM classifiers achieves a similar accuracy to a homogeneous solely based on Hoeffding Trees or Naive Bayes we have evaluated configurations of PDM that use both classifiers. Again all experiments have been conducted five times and the average of the achieved accuracy by the MADM has been calculated.

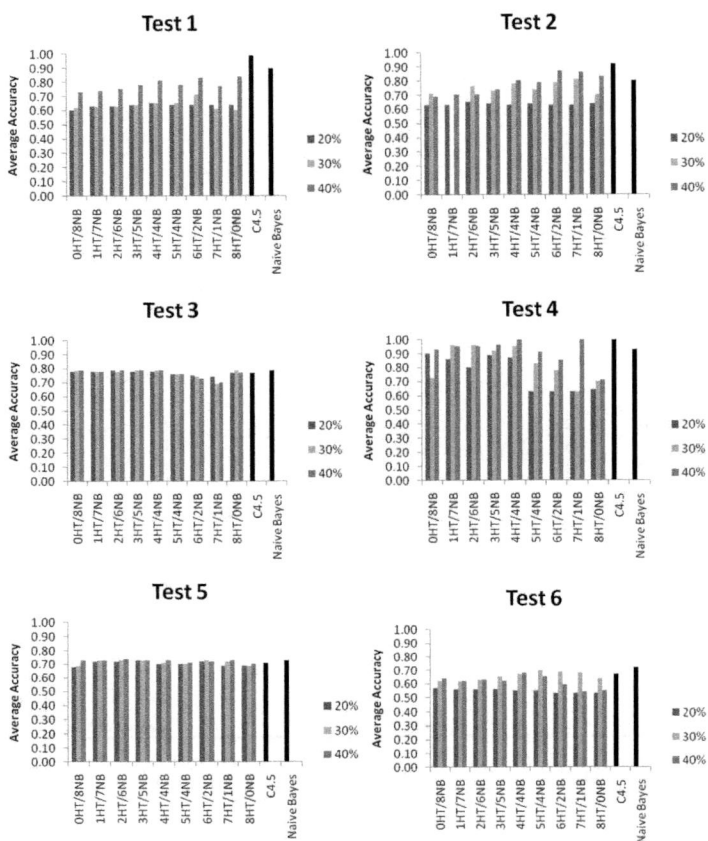

Fig. 7. Average classification accuracy for a heterogeneous set-up of classifier AMs in PDM

Figure 7 highlights experiments conducted with different heterogeneous setups of PDM for all 6 datasets listed in Table 1. The average accuracy of the MADM is plotted against the combination of algorithms embedded in the deployed AMs. The graph is split showing AMs working with 20%, 30% and 40% of the features. All possible combinations of Hoeffding Trees and Naive Bayes AMs have been evaluated. The horizontal labels in Figure 7 are read the following way. HT stands for Hoeffding Tree and NB for Naive Bayes, the number before HT and NB is the number of Naive Bayes or HT classifiers visited respectively. For example label '3HT/5NB' means that 3 Hoeffding Tree AMs and 5 NB agents have been visited by the MADM. Also plotted in Figure 7 is the result the batch learning algorithms C4.5 and Naive Bayes achieve using all the features. The achieved accuracies are close compared with those achieved by the batch learning algorithms which have the advantage over *PDM* of having all the features available which would again not be the case in a realistic scenario where subscribers of a data stream limit their subscription only to properties they are particularly interested in for reasons stated in Sections 1 and 4.

In Figure 7 for test 1 it can be seen that using 20% features or 30% seems to achieve very similar classification accuracies. However for using 40% features the classification accuracy improves considerably and gets close to the batch learning accuracies which use all features, also for using 40% of the features it can be seen that configurations with more Hoeffding tree AMs perform slightly better than configurations with more Naive Bayes AMs. For test 2 it can be seen that using 30% instances already improves the classification accuracy and there is a tendency for the usage of 30% and 40% instances that configurations with more Hoeffding tree AMs perform slightly better, in particular configurations 6HT/2NB and 7HT/1NB seem to achieve high accuracies between 80% and 90%. Regarding test 3 all percentages of features used achieve a very similar and very good classification accuracies that can well compete with the accuracies achieved by the batch learning algorithms on all features. Also configuration wise it seems that using configurations with more Naive Bayes then Hoeffding trees seem to perform better, however also a configuration solely based on Hoeffding trees achieves very good classification accuracies close to batch learning algorithms. In test 4 the tendency seems that using more Naive Bayes classifiers achieve a higher accuracy than using more Hoeffding tree classifiers. In test 4 Naive Bayes seems to work better on configurations with less features subscribed to compared with Hoeffding trees. There is no noticeable tendency for test 5, all configurations and percentages of features seem to achieve a high accuracy very close to the one observed for the batch learning algorithms. On test 6 any configuration and even batch learning algorithms do not achieve a good classification accuracy. This suggests that the dataset for test 6 is not very well suited for classification in its current form. Also there is no particular tendency detectable for test 6.

Figure 7 also displays the data from homogeneous configurations of PDM solely based on Hoeffding trees, which is labelled as configuration 8HT/0NB and solely based on Naive Bayes, which is labelled 0HT/8NB. The results clearly show that heterogeneous configurations of PDM achieve very similar accuracies to homogeneous configurations of PDM. Also earlier in this section we stated that a heterogeneous set-up of AM classifiers would also be a more realistic setup as owners of mobile devices may use their individual classification techniques. Furthermore PDM may well benefit from using individual AMs from different owners as they are likely to be optimised on the local subscription of the data stream.

5   Ongoing and Future Work

5.1   Rating System for AMs

The current implementation of the MADM agent assumes that the local AMs are of good quality and thus in the case of classification of unlabelled data instances it is assumed that the weights are calculated correctly and truly reflect the AMs classification accuracy. This assumption may be true for the AMs we developed in-house, which we used for the evaluation in Section 4, but third party implementations may not be trusted. For this reason a rating system about AMs is currently being developed based on historical consultations of AMs by the

MADM. For example if the MADM remembers the classifications and weights obtained from AMs visited, and the true classification of the previously unknown instances is revealed, then the MADM could implement its own rating system and rate how reliable a AM's weight was in the past. If an AM is rated as unreliable, then the MADM may even further lower its weight. However it is essential that this rating system is also able to loosen given ratings, as the AM's performance might well change if there is a concept drift in the data stream. In order to detect such concept drifts it is necessary that AMs that have a bad rating are still taken into consideration, even if it is with a low impact due to bad ratings.

5.2   Intelligent Schedule for MADMs

In its current implementation the MADM visits all available AMs, however this may be impracticable if the number of AMs is very large. Currently a mechanism is being developed for MADMs according to which the MADM can decide when to stop consulting further AMs. A possible stopping criteria could be that a certain time has elapsed or the classification result is reliable enough. Also the rating system outlined above can be used to determine an order in which AMs are visited. If there are time constraints the MADM may prioritise more reliable AMs.

6   Conclusions

The paper presents the Pocket Data Mining (PDM) framework for mining data streams in a collaborative fashion in a mobile environment consisting of smart phones. PDM is based on mobile agent technology using three types of agents. Agent Miners that embed stream mining technologies, Mobile Agent Resource Discoverers that are used to roam the network and search for relevant data streams and the Mobile Agent Decision Makers that visit and consults Agent Miner on whose results they base their final decision on. This paper presents a implementation of PDM for distributed classification of data streams and examines its feasibility. Two different configurations of PDM based on either Hoeffding tree Agent Miners or Naive Bayes Agent Miners have been examined. The experiments have also been conducted with different percentages of features available to the Agent Miners. The classifiers used were the Hoeffding tree and the Naive Bayes classifiers. In general it has been observed that any configuration of PDM based on any of the two or both classifiers achieved an acceptable classification accuracy. Also it has been observed that the more features the Agent Miners have available the closer the accuracies are to the ideal case where batch learning algorithms C4.5 and Naive Bayes have the advantage of having 100% of the features available to train the classifier. Furthermore it has been observed that PDM's 'weighted' majority voting achieves higher classification accuracies than simply taking the Agent Miners local average accuracies. In general it does not seem that one of the classifiers is superior in general, this indicates that heterogeneous configurations of PDM using different classifiers in the same network will perform equally well. Also this would be the more realistic scenario as owners of smart phones are likely to employ the classifiers that are likely to perform well on their data subscription. To examine this further a heterogeneous setup of PDM using both Naive Bayes and Hoeffding tree classifiers have been evaluated. Again the classification accuracies were very similar to the ones achieved with homogeneous setups of PDM.

Ongoing work comprises a rating system to rate the quality of third party AMs; the development of an optimised schedule for MADM agents in order to derive data mining results faster if there are many available AMs.

PDM is a new niche of distributed data mining however hardly explored. The current implementation of PDM focuses on classification techniques, however, there exist many more data mining technologies tailored for data streams and mobile devices. For example there are stream mining techniques that classify unlabelled data streams [30,32] which could be introduced into PDM. Also resource aware data mining algorithms as proposed in [29] will boost PDM's applicability in resource constraint mobile networks once integrated in PDM. But not only data streams can be used also the mobile phones sensors such as the gyroscope, camera, etc. could be used as data sources as well. In general PDM's applicability will benefit with the recent advances in smart phone technology and data stream mining technology and vice versa.

References

1. Aggarwal, C.C., Han, J., Wang, J., Yu, P.: A Framework for Clustering Evolving Data Streams. In: Proceedings of the VLDB Conference (2003)
2. Aggarwal, C.C., Han, J., Wang, J., Yu, P.: On Demand Classification of Data Streams. In: Proceedings of the ACM KDD Conference (2004)
3. Stahl, F., Gaber, M.M., Bramer, M., Yu, P.S.: Pocket Data Mining: Towards Collaborative Data Mining in Mobile Computing Environments. In: Proceedings of the IEEE 22nd International Conference on Tools with Artificial Intelligence (ICTAI 2010), Arras, France, October 27-29 (2010)
4. Bifet, A., Holmes, G., Pfahringer, B., Kranen, P., Kremer, H., Jansen, T., Seidl, T.: Journal of Machine Learning Research, JMLR (2010)
5. Bifet, A., Kirkby, R.: Data Stream Mining: A Practical Approach, Center for Open Source Innovation (August 2009)
6. Gaber, M.M., Zaslavsky, A., Krishnaswamy, S.: Mining Data Streams: A Review. ACM SIGMOD Record 34(1), 18–26 (2005) ISSN: 0163-5808
7. Zaslavsky, A.: Mobile Agents: Can They Assist with Context Awareness? In: IEEE MDM, Berkeley, California (January 2004)
8. Page, J., Padovitz, A., Gaber, M.: Mobility in Agents, a Stumbling or a Building Block? In: Proceedings of Second International Conference on Intelligent Computing and Information Systems, Cairo, Egypt, March 5-7 (2005)
9. da Silva, J., Giannella, C., Bhargava, R., Kargupta, H., Klusch, M.: Distributed Data Mining and Agents. Engineering Applications of Artificial Intelligence Journal 18, 791–807 (2005)

10. Kargupta, H., Hamzaoglu, I., Stafford, B.: Scalable, Distributed Data Mining Using an Agent-Based Architecture. In: Heckerman, D., Mannila, H., Pregibon, D., Uthurusamy, R. (eds.) Proceedings of Knowledge Discovery and Data Mining, pp. 211–214. AAAI Press (1997)

11. Pittie, S., Kargupta, H., Park, B.: Dependency Detection in MobiMine: A Systems Perspective. Information Sciences Journal 55(3-4), 227–243 (2003)

12. Krishnaswamy, S., Loke, S.W., Zaslavsky, A.B.: A hybrid model for improving response time in distributed data mining. IEEE Transactions on Systems, Man, and Cybernetics, Part B 34(6), 2466–2479 (2004)

13. Domingos, P., Hulten, G.: Mining high-speed data streams. In: International Conference on Knowledge Discovery and Data Mining, pp. 71–80 (2000)

14. Langley, P., Iba, W., Thompson, K.: An analysis of bayesian classifiers. In: National Conference on Artificial Intelligence, pp. 223–228 (1992)

15. Witten, I., Frank, E.: Data Mining: Practical Machine Learning Tools and Techniques with Java Implementations, 2nd edn. Morgan Kaufmann (2005)

16. Bellifemine, F., Poggi, A., Rimassa, G.: Developing Multi-Agent Systems with JADE. In: Castelfranchi, C., Lespérance, Y. (eds.) ATAL 2000. LNCS (LNAI), vol. 1986, pp. 89–103. Springer, Heidelberg (2001)

17. Blake, C.L., Merz, C.J.: UCI Repository of Machine Learning Databases (Technical Report). University of California, Irvine, Department of Information and Computer Sciences (1998)

18. Bacardit, J., Krasnogor, N.: The Infobiotics, PSP benchmarks repository (2008), http://www.infobiotic.net/PSPbenchmarks

19. Kargupta, H., Park, B., Pittie, S., Liu, L., Kushraj, D., Sarkar, K.: MobiMine: Monitoring the Stock Market from a PDA. ACM SIGKDD Explorations 3(2), 37–46 (2002)

20. Kargupta, H., Bhargava, R., Liu, K., Powers, M., Blair, P., Bushra, S., Dull, J., Sarkar, K., Klein, M., Vasa, M., Handy, D.: VEDAS: A Mobile and Distributed Data Stream Mining System for Real-Time Vehicle Monitoring. In: Proceedings of the SIAM International Data Mining Conference, Orlando (2004)

21. Kargupta, H., Puttagunta, V., Klein, M., Sarkar, K.: On-board Vehicle Data Stream Monitoring using MineFleet and Fast Resource Constrained Monitoring of Correlation Matrices. Next Generation Computing. Invited Submission for Special Issue on Learning from Data Streams 25(1), 5–32 (2007)

22. Park, B., Kargupta, H.: Distributed Data Mining: Algorithms, Systems, and Applications. In: Ye, N. (ed.) Data Mining Handbook (2002)

23. Agnik, MineFleet Description, http://www.agnik.com/minefleet.html

24. Quinlan, J.R.: C4.5: Programs for Machine Learning. Morgan Kaufmann Publishers (1993)

25. Krishnaswamy, S., Gaber, M.M., Harbach, M., Hugues, C., Sinha, A., Gillick, B., Haghighi, P.D., Zaslavsky, A.: Open Mobile Miner: A Toolkit for Mobile Data Stream Mining. In: Proceedings of the 15th ACM SIGKDD International Conference on Knowledge Discovery and Data Mining 2009, Paris, France, June 28-1 July (2009) (Demo paper)

26. BBC, Budget Cuts of Police Force, http://www.bbc.co.uk/news/uk-10639938

27. Poh, M., Kim, K., Goessling, A.D., Swenson, N.C., Picard, R.W.: Heartphones: Sensor Earphones and Mobile Application for Non-obtrusive Health Monitoring. In: IEEE International Symposium on Wearable Computers, Austria, pp. 153–154 (2009)

28. Gaber, M.M., Zaslavsky, A.B., Krishnaswamy, S.: Data Stream Mining. In: Data Mining and Knowledge Discovery Handbook 2010, pp. 759–787. Springer, Heidelberg (2010)

29. Gaber, M.M., Krishnaswamy, S., Zaslavsky, A.: Resource-Aware Mining of Data Streams. Journal of Universal Computer Science 11(8), 1440–1453 (2005) ISSN 0948-695x, Special Issue on Knowledge Discovery in Data Streams, Verlag der Technischen Universit Graz, Know-Center Graz, Austria (August 2005)

30. Gaber, M.M., Yu, P.S.: A framework for resource-aware knowledge discovery in data streams: a holistic approach with its application to clustering. In: Proceedings of the 2006 ACM Symposium on Applied Computing (SAC), Dijon, France, April 23-27, pp. 649–656. ACM Press (2006)

31. Gaber, M.M.: Data Stream Mining Using Granularity-Based Approach. In: Foundations of Computational Intelligence, vol. (6), pp. 47–66. Springer, Heidelberg (2009)

32. Phung, N.D., Gaber, M.M., Ohm, U.R.: Resource-aware online data mining in wireless sensor networks. In: Proceedings of the IEEE Symposium on Computational Intelligence and Data Mining (CIDM 2007), April 1-5 (2007)

Integrated Distributed/Mobile Logistics Management

Lars Frank and Rasmus Ulslev Pedersen

Department of IT Management, Copenhagen Business School, Howitzvej 60,
DK-2000 Frederiksberg, Denmark
lpf.inf@cbs.dk

Abstract. The objective of the logistics management and control in a transport enterprise is to plan for the cheapest way to fulfill the transport needs of the customers and to offer services to the customers like for example supplying information about where products of the customers are in the transport process at any time. In case of delays in some transports this type of information may be important if the customers have to fulfill their obligations. The objective of this paper is to describe an architecture for logistics management systems where it is possible to integrate the logistics management systems of different cooperating transport companies or mobile users in order to optimize the availability to data that can optimize the transport process. In central databases the consistency of data is normally implemented by using the ACID (Atomicity, Consistency, Isolation and Durability) properties of a DBMS (Data Base Management System). This is not possible if distributed and/or mobile databases are involved and the availability of data also has to be optimized. Therefore, we will in this paper use so-called relaxed ACID properties across different locations. The objective of designing relaxed ACID properties across different database locations is that the users can trust the data they use even if the distributed database is temporarily inconsistent. It is also important that disconnected locations can operate in a meaningful way in so-called disconnected mode.

Keywords: Logistics management, ERP systems, mobility, replication methods, relaxed ACID properties, database availability, fault tolerance, multidatabases.

1   Introduction

In large transport companies, logistics are normally managed by using specialized ERP (Enterprise Resource Planning) modules that use a central database common for the company. When using a DBMS to manage the database an important aspect involves transactions which are any logical operation on data and per definition database transactions must be atomic, consistent, isolated, and durable in order for the transaction to be reliable and coherent.

The ACID properties of a database are delivered by a DBMS to make database recovery easier and make it possible in a multi-user environment to give concurrent transactions a consistent chronological view of the data in the database. Some challenges associated with this are acknowledge by [17]. The ACID properties are consequently important for users that need a consistent view of the data in a database.

A. Hameurlain et al. (Eds.): TLDKS V, LNCS 7100, pp. 206–221, 2012.

However, the implementation of ACID properties may influence performance and slow down the availability of a system in order to guarantee that all users have a consistent view of data even in case of failures. The challenges and tradeoffs of global transaction management are also outlined by Sheth and Larson [27]. The challenge of high availability is well recognized in the Internet setting where poor response time is linked to profit losses. Loukopoulos, Ahmad, and Papadias [20] express it in this way:

> "The boost in the popularity of the web resulted in large bandwidth demands and notoriously high latencies experienced by end-users. To combat these problems the benefits of caching and replication were early recognized by the research community."

In several situations, the availability and the response time will be unacceptable if the ACID properties of a DBMS are used without reflection. Pre-serilization is one possible counter measure for this challenge [25]. This is especially the case in distributed and/or mobile databases [23] where a failure in the connections of a system should not prevent the system from operating in a meaningful way in disconnected mode.

Information systems that operate in different locations can be integrated by using more or less common data and/or by exchanging information between the systems involved. In both situations, the union of the databases of the different systems may be implemented as a database with so called relaxed ACID properties where temporary inconsistencies may occur in a controlled manner. However, when implementing relaxed ACID properties it is important that from a user's point of view it must still seem as if traditional ACID properties were implemented, which therefore will keep the local databases trustworthy for decision making.

In the following part of the introduction, we will give an overview of how relaxed ACID properties may be implemented and used in central and mobile databases integrated by using relaxed ACID properties. Next, we will give a more detailed description of the objectives of the paper.

1.1   Relaxed ACID Properties

The Atomicity property of a DBMS guarantees that either all the updates of a transaction are committed/executed or no updates are committed/executed. This property makes it possible to re-execute a transaction that has failed after execution of some of its updates. In distributed databases, this property is especially important if data are replicated as inconsistency will occur if only a subset of data is replicated. The Atomicity property of a DBMS is implemented by using a DBMS log file with all the database changes made by the transactions. The global Atomicity property of databases with relaxed ACID properties is implemented by using compensatable, pivot and retriable subtransactions in that sequence as explained in section 2.1. By applying these subtransactions it is allowed to commit/execute only part of the transaction and still consider the transaction to be atomic as the data converge towards a consistent state.

As explained in section 2.2 the global Consistency property is not defined in databases with relaxed ACID properties because normally such databases are inconsistent and this inconsistency may be managed in the same way as the relaxed Isolation property.

The Isolation property of a DBMS guarantees that the updates of a transaction cannot be seen by other concurrent transactions until the transaction is committed/executed. That is the inconsistencies caused by a transaction that has not executed all its updates cannot be seen by other transactions. The Isolation property of a DBMS may be implemented by locking all records used by a transaction. That is the locked records cannot be used by other transactions before the locks are released when the transaction is committed. The global Isolation property of databases with relaxed ACID properties is implemented by using countermeasures against the inconsistencies/anomalies that may occur. This is explained in more details in section 2.3.

The Durability property of a DBMS guarantees that the updates of a transaction cannot be lost if the transaction is committed. The Durability property of a DBMS is implemented by using a DBMS log file with all the database changes made by the transactions. By restoring the updates of the committed transactions it is possible to recover a database even in case it is destroyed. The global Durability property of databases with relaxed ACID properties is implemented by using the local Durability property of the local databases involved.

Data replication is normally used to decrease local response time and increase local performance by substituting remote data accesses with local data accesses [7]. At the same time, the availability of data will normally also be increased as data may be stored in all the locations where they are vital for disconnected operation. These properties are especially important in mobile applications where the mobile user often may be disconnected from data that are vital for the normal operation of the application [18]. See [3] for a framework that addressee such challenges. The major disadvantages of data replication are the additional costs of updating replicated data and the problems related to managing the consistency of the replicated data.

In general, replication methods involve n copies of some data where n must be greater than 1. The basic replication designs storing n different copies of some data are defined to be n-safe, 2-safe, 1-safe or 0-safe, respectively, when n, 2, 1 or 0 of the n copies are consistent and up-to-date at normal operation. An overview of replication in distributed systems can be found in [28]. In mobile computing, we recommend using only the 1-safe and 0-safe replication designs, as these use asynchronous updates which allow local system to operate in disconnected mode.

In this paper, we will use design rules for selecting replication designs that make it easier to implement countermeasures against isolation anomalies.

1.2   Objective

The objective of this paper is to describe an architecture for logistics management systems where it is possible to integrate the logistics management systems of different cooperating transport companies and mobile users in order to make it possible to optimize the integrated logistics management and make the integrated system flexible to unplanned changes in transports and the transported objects. It is also an objective that the different local systems may be heterogeneous as different transport operators may not have the same type of logistics management systems. For all mobile systems it is an objective that it should be possible for them to operate in disconnected mode as failures are more common in mobile connections. However, it is also a design objective that the stationary logistics systems can operate in disconnected mode as they may belong to different Transport operators.

The paper is organized as follows: First, we will describe the transaction model used in this paper. Applications may often be optimized and made fault tolerant by using replicated data. Therefore, we will in section 3 describe how it is possible with different levels of relaxed ACID properties to asynchronously replicate data from one local database to another. In section 4, we will in details describe how it is possible in an application like logistic management and control to select the replication designs that optimize the application. Concluding remarks and future research are presented in section 5.

Related Research: The transaction model used in this paper is *the countermeasure transaction model* [6]. This model uses the relaxed Atomicity property as developed by [12], [21], [30] and [31]. How to use countermeasures against consistency anomalies are described by [10]. The model uses countermeasures against the isolation anomalies. These anomalies are described by [1] and [2]. How to use countermeasures against isolation anomalies was first systematized by [11]. In the countermeasure transaction model the problem of implementing the Durability property is solved as describe by [2]. Like [31], we might call our transactions *'flexible transactions'* as the only new facility is the use of countermeasures to manage the missing consistency and isolation properties.

A lot of research in Distributed Data Base Management Systems (DDBMS) has taken place for many years. However, this research has not had practical application in mobile computing where it is an objective that mobile users should have the possibility to operate in disconnected mode. Anyway, the location-transparency of a DDBMS query [16] and [29] can be adopted in the countermeasure transaction model and may be useful in many distributed applications. However, in the application of this paper we will not recommend this possibility as we use asynchronous data replication in order to make it possible for the mobile users to operate in disconnected mode. The replication designs used in this paper are described in more detail by [4][7]. The design rules used for choosing among the different replication designs are described by [10].

This paper is a significant improved version of [6] as structural errors in the logistics management database of figure 3 and misunderstandings in logistic management has been corrected.

2   The Transaction Model

A *multidatabase* is a union of local autonomous databases. *Global transactions* [14] access data located in more than one local database. In recent years, many transaction models have been designed to integrate local databases without using a distributed DBMS. The countermeasure transaction model [11] has, among other things, selected and integrated properties from these transaction models to reduce the problems caused by the missing ACID properties in a distributed database that is not managed by a distributed DBMS. In the countermeasure transaction model, a global transaction involves a *root transaction* (client transaction) and several single site *subtransactions* (server transactions). Subtransactions may be nested transactions, i.e. a subtransaction may be a *parent transaction* for other subtransactions. All communication with the user is managed from the root transaction, and all data is accessed through subtransactions. The following subsections will give a broad outline of how relaxed ACID properties are implemented.

2.1   The Atomicity Property

An updating transaction has the *atomicity property* and is called *atomic* if either all or none of its updates are executed. In the countermeasure transaction model, the global transaction is partitioned into the following types of subtransactions executed in different locations:

The *pivot* subtransaction that manages the atomicity of the global transaction. The global transaction is committed when the pivot subtransaction is committed locally. If the pivot subtransaction aborts, all the updates of the other subtransactions must be compensated.

The *compensatable* subtransactions that all may be compensated. Compensatable subtransactions must always be executed before the pivot subtransaction is executed to make it possible to compensate them if the pivot subtransaction cannot be committed. A compensatable subtransaction may be compensated by executing a *compensating* subtransaction. Compensation mechanisms are also discussed in relation to local transaction managers [32].

The *retriable* subtransactions that are designed in such a way that the execution is guaranteed to commit locally (sooner or later) if the pivot subtransaction has been committed.

The global atomicity property is implemented by executing the compensatable, pivot and retriable subtransactions of a global transaction in that order. For example, if the global transaction fails before the pivot has been committed, it is possible to remove the updates of the global transaction by compensation. If the global transaction fails after the pivot has been committed, the remaining retriable subtransactions will be (re)executed automatically until all the updates of the global transaction have been committed.

2.2   The Consistency Property

A database is *consistent* if its data complies with the consistency rules of the database. If the database is consistent both when a transaction starts and when it has been completed and committed, the execution has the *consistency property*. Transaction *consistency rules* may be implemented as a control program that rejects the commitment of transactions, which do not comply with the consistency rules.

The above definition of the consistency property is not useful in distributed databases with relaxed ACID properties because such a database is almost always inconsistent. However, a distributed database with relaxed ACID properties should have *asymptotic consistency*, i.e. the database should converge towards a consistent state when all active transactions have been committed/compensated. Therefore, the following property is essential in distributed databases with relaxed ACID properties:

> If the database is asymptotically consistent when a transaction starts and also when it has been committed, the execution has the *relaxed consistency property*.

2.3   The Isolation Property

The isolation property is normally implemented by using *long duration locks*, which are locks that are held until the global transaction has been committed [11. In the countermeasure transaction model, long duration locks cannot instigate isolated global execution as retriable subtransactions may be executed after the global transaction has been committed in the pivot location. Therefore, *short duration locks* are used, i.e. locks that are released immediately after a subtransaction has been committed/aborted locally. To ensure high availability in locked data, short duration locks should also be used in compensatable subtransactions, just as locks should be released before interaction with a user. This is not a problem in the countermeasure transaction model as the traditional isolation property in retriable subtransactions is lost anyway. If only short duration locks are used, it is impossible to block data. (Data is *blocked* if it is locked by a subtransaction that loses the connection to the "coordinator" (the pivot subtransaction) managing the global commit/abort decision). When transactions are executed without isolation, the so-called *isolation anomalies* may occur. In the countermeasure transaction model, relaxed isolation can be implemented by using countermeasures against the isolation anomalies. If there is no isolation and the atomicity property is implemented, the following isolation anomalies may occur [1][2].

- *The lost update anomaly* is by definition a situation where a first transaction reads a record for update without using locks. Subsequently, the record is updated by another transaction. Later, the update is overwritten by the first transaction. In extended transaction models, the lost update anomaly may be prevented, if the first transaction reads and updates the record in the same subtransaction using local ACID properties. Unfortunately, the read and the update are sometimes executed in different subtransactions belonging to the same parent transaction. In such a situation, a second transaction may update the record between the read and the update of the first transaction.
- *The dirty read anomaly* is by definition a situation where a first transaction updates a record without committing the update. Subsequently, a second transaction reads the record. Later, the first update is aborted (or committed), i.e. the second transaction may have read a non-existing version of the record. In extended transaction models, this may happen when the first transaction updates a record by using a compensatable subtransaction and later aborts the update by using a compensating subtransaction. If a second transaction reads the record before it has been compensated, the data read will be "dirty".

- *The non-repeatable read anomaly* or *fuzzy read* is by definition a situation where a first transaction reads a record without using locks. Later, the record is updated and committed by a second transaction before the first transaction has been committed. In other words, it is not possible to rely on the data that have been read. In extended transaction models, this may happen when the first transaction reads a record that later is updated by a second transaction, which commits the update locally before the first transaction commits globally.
- *The phantom anomaly* is by definition a situation where a first transaction reads some records by using a search condition. Subsequently, a second transaction updates the database in such a way that the result of the search condition is changed. In other words, the first transaction cannot repeat the search without changing the result. Using a data warehouse may often solve the problems of this anomaly [6].

The countermeasure transaction model [11] describes countermeasures that eliminate or reduce the problems of the isolation anomalies. In this paper, it is assumed that the following countermeasures are used:

The *reread countermeasure* is primarily used to prevent the lost update anomaly. Transactions that use this countermeasure read a record twice using short duration locks for each read. If a second transaction has changed the record between the two readings, the transaction aborts itself after the second read. In the replicated databases used in this paper the reread countermeasure will only function if there is a primary copy location.

The *Commutative Updates Countermeasure* can prevent lost updates merely by using commutative operators. Adding and subtracting an amount from an account are examples of commutative updates. If a subtransaction only has commutative updates, it may be designed as commutable with other subtransactions that only have commutative updates as well. In the replicated databases used in this paper the Commutative Updates countermeasure is recommended when there is no primary copy location.

The *Pessimistic View Countermeasure* reduces or eliminates the dirty read anomaly and/or the non-repeatable read anomaly by giving the users a pessimistic view of the situation. In other words, the user cannot misuse the information. The purpose is to eliminate the risk involved in using data where long duration locks should have been used. The pessimistic view countermeasure may be implemented by using:

- The pivot or compensatable subtransactions for updates that "limit" the users' options. That is, concurrent transactions cannot use resources that are reserved for the compensatable/pivot subtransaction.
- The pivot or retriable subtransactions for updates that "increase" the users' options. That is, concurrent transactions can only use increased resources after the increase has been committed.

2.4   The Durability Property

Updates of transactions are said to be *durable* if they are stored in a stable manner and secured by a log recovery system. In case a global transaction has the atomicity property (or relaxed atomicity), the global durability property (or relaxed durability property) will automatically be implemented, as it is ensured by the log-system of the local DBMS systems [2].

3   Description of the Most Important Asynchronous Replication Methods

The n-safe and quorum safe replication designs are not suited for systems where it should be possible to operate in disconnected mode. Therefore, we will only describe the asynchronous replication methods in the following. For all these replication methods it is possible to implement them by using compensatable and retriable services from other locations.

3.1   The Basic 1-Safe Design

In the basic 1-safe design [14], the primary transaction manager goes through the standard commit logic and declares completion when the commit record has been written to the local log. In the basic 1-safe design, the log records are asynchronously spooled to the locations of the secondary copies. In case of a primary site failure in the basic 1-safe design, production may continue by selecting one of the secondary copies but this is not recommended in distributed ERP systems as it may result in *lost transactions*.

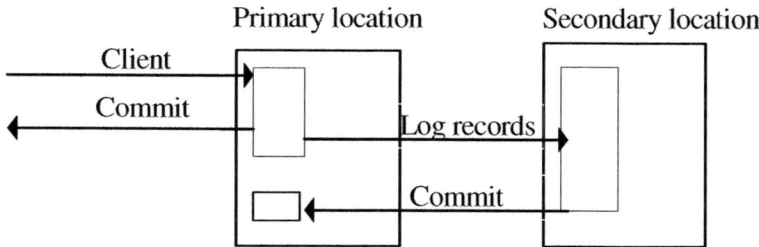

Fig. 1. The basic 1-safe database design

3.2   The 0-Safe Design with Local Commit

The *0-safe design with local commit* is defined as n table copies in different locations where each transaction first will go to the nearest database location, where it is executed and committed locally.

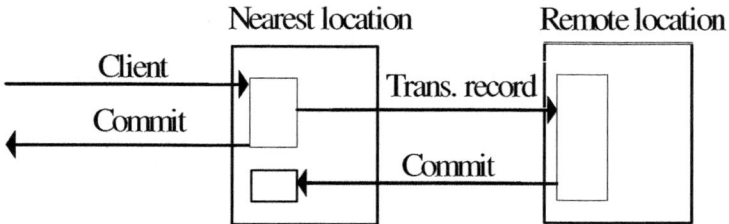

Fig. 2. The 0-safe database design

If the transaction is an update transaction, the transaction propagates asynchronously to the other database locations, where the transaction is re-executed without user dialog and committed locally at each location. This means that all the table copies normally are inconsistent and not up to date under normal operation. The inconsistency must be managed by using countermeasures against the isolation anomalies. For example, to prevent lost updates in the 0-safe design, all update transactions must be designed to be commutative [11].

3.3   The 1-Safe Design with Commutative Updates

The 1-safe design can transfer updates to the secondary copies in the same way as the 0-safe design. In such a design, lost transactions cannot occur because the transaction transfer of the 0-safe design makes the updates commutative. The properties of this mixed design will come from either the basic 1-safe or the 0-safe design. The 1-safe design with commutative updates does not have the high update performance and capacity of the 0-safe design. On the other hand, in this design the isolation property may be implemented automatically as long as the primary copy does not fail. Normally, this makes it much cheaper to implement countermeasures against the isolation anomalies, because it is only necessary to secure that "the lost transactions" are not lost in case of a primary copy failure.

3.4   The 0-Safe Design with Primary Copy Commit

The *"0-safe designs with primary copy commit"* is a replication methods where a global transaction first must update the local copy closest to the user by using a compensatable subtransaction. Later, the local update may be committed globally by using a primary copy location. If this is not possible, the first update must be compensated. The primary copy may have a version number that also are replicated. This may be used to control that an updating user operates on the latest version of the primary copy before an update is committed globally in the location of the primary copy. If an update cannot be committed globally in the primary copy location, the latest version of the primary copy should be send back to the updating user in order to repeat the update by using the latest version of the primary copy.

3.5   Implementation of Internet Replication Services with Relaxed ACID Properties

In order to implement integration flexibility (fault tolerance in case a backup unit is used) between the different ERP modules it is not acceptable that the different modules communicate directly with each other.

All communication between different modules must be executed by applications offered as e.g. SOA (Service Oriented Architecture) services. The use of application services for making information transfers between different ERP modules make it possible to integrate heterogeneous ERP modules developed by different software suppliers. In order also to implement the atomicity property between the modules each module should offer the following types of services [8]:

- Read only services that are used when a system unit wants to read data managed by another system.
- Compensatable update services that are used when a system unit wants to make compensatable updates in tables managed by another system.
- Retriable update services that are used when a system unit wants to make retriable updates in tables managed by another system.

The SOA services may be used to implement the asynchronous replication designs in the following way [10]:

- All types of the 1-safe designs can be implemented in the following way: Either the primary copy location uses the retriable services of the secondary copy locations directly to transfer the replicated records, or the primary copy location can use the retriable service of a distribution center to transfer the replicated records indirectly. The first solution with direct replication transfer is fast but the solution implies that each primary copy location knows how to contact the different secondary locations. In the solution with indirect transfers, it is only the distribution center that should know how to contact the different secondary copy locations.
- The 0-safe design with local commit can be implemented in the same ways as the 1-safe designs except that all locations should function as a primary copy location when they have records to replicate.
- The 0-safe design with primary copy commit can be implemented in the following way: First a compensatable update is executed locally in the initiating location. Next, a retriable service call to a primary copy location is executed, and if the update is committed the primary copy location can transfer the replicated records in the same ways as the 1-safe designs. However, if the update is rejected in the primary copy location, a retriable service call is executed in order to remove the original update from the initiating location.

4   Implementation of an Integrated System for Mobile Logistics Management and Control

The following figure illustrates the most important tables in a system for logistics management and control in a transport enterprise that receives transport Orders from their Customers. The importance of such a real-time tracking systems is underpinned by [13]. An overview from a Supply Chain perspective can be found [24]. A more general ER model can be found in the open source ERP system OpenBravo [15][26].

In this case, we assume that all packages covered by a transport Order are sent from one location to another by the customer's choice. The objective of the logistics management and control is to plan for the most economical way to fulfill the transport needs of the customers. In case of damages to the customers' packages or delays in their transports, it is also very important that the planned transports may be changed dynamically and that it is possible for the customers to intervene in order for them to fulfill their obligations in the best possible way.

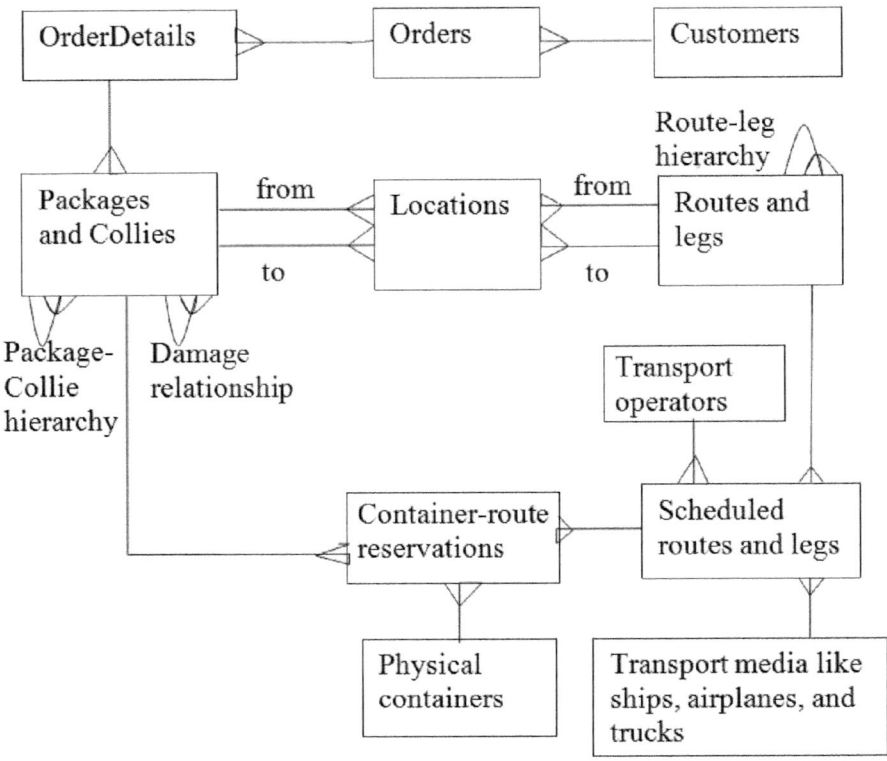

Fig. 3. ER-diagram of the most important tables in a logistics management system

Most of the tables illustrated in Figure 3 are fragmented, as different users need to know only about the packages, collies, transports, etc., that they are involved in. However, another reason for fragmentation is that different fragments of the tables may be replicated by using different replication methods, as will be illustrated later. A transport enterprise may use several different ERP systems, as it may have different centers around the world. In this paper, we assume that only one ERP system is used by the primary Transport operator because different centers with ERP systems of their own may be viewed as foreign Transport operators. A stationary database server belonging to the primary Transport operator may function as backup and recovery system for the tables of the mobile users. The stationary database server can be used to store all retriable updates used by the different replication designs, as described in section 2.

In Fig. 3, the Orders of the Customers may contain many Order Details, and an Order Detail may consist of many Packages that are going to the same location.

Package and OrderDetail records are created in the order entry module of an ERP system. Packages that have one or more Route destinations in common are stored in 'so called' Collies. A Collie is normally packed on a pallet in order to make it easier to move the Collie, for example by truck. Both Packages and Collies may be damaged. If a Package is damaged, it may be split into more Packages with relationship to the parent Package for which the papers for the Custom Authorities are made. If a Collie is damaged, it may be repacked without restrictions, as long as the relationship table "Collies-packages" describes in which Collie a given Packet is stored at any given time. Collies are either under transport or at its departure or arrival location.

A Route may be viewed as a transport channel consisting of an ordered set of so called Legs, where each Leg has a departure Location and a destination Location without intermediate stops where the Collies can be redirected to new destinations. A Collie can be transported via many Routes from its first departure to its final destination, as long as the new Route starts where the previous one ended. The table "Scheduled routes and legs" has a record for each scheduled transport through a Route or Leg. The Container-route reservations keeps track on which Scheduled Route a given Container is (planned to be) transported at any given time. The Transport operators are the responsible owners of the scheduled routes stored in the table "Scheduled routes and legs". The enterprise for which the ER-diagram is designed is a Transport operator, but no operator can cover all routes at all times and therefore, the ER-diagram has the possibility to operate with many different Transport operators. It is thus possible to use foreign operators as sub-suppliers for transport services when the Primary Transport operator cannot cover a route in a proper way.

In the rest of this section we will describe and argue for the replication designs we will recommend for the database tables in our logistics architecture.

OrderDetail, Package , Collie and Container-route reservation records may be created and replicated by using the basic 1-safe design with the primary copy in the order entry ERP system of the primary Transport operator and secondary copies in all the mobile and de-central locations to which the packages are planned to travel. However, in practice the primary copy property should follow the packages and in this way optimize possible updates as the only ones that know the exact state of the Packages and Collies are the current Operator of the Packages/Collies/Container-route reservations and their possible mobile controllers that are responsible for controlling arrival and departure of physical containers/collies/packages. Therefore, we will recommend that Packages, Collies and their related Container-route reservations are replicated by using the 0-safe design with primary copy commit in the ERP system of the current Operator that for the time being has the responsibility for Package/Collie. It is also possible to use 0-safe design with local commit as "lost updates" normally not are a serious problem because all logistics updates are either record creations or attribute substitutes that should be preserved in the historic versions and not overwritten.

We recommend that the tables with "Routes and legs" and "Scheduled routes and legs" is fragmented and replicated by using the basic 1-safe design with the primary copy in the ERP system of the Operator that operate the Route/Leg and secondary copies in all the mobile and de-central controllers that may use the Route/Leg. By using these tables, it is possible for any mobile user to make the decision to change the Route of a collie/container in order to adapt for non-planned changes.

We recommend that the tables with Transport operators, Transport media, and Locations are replicated by using the basic 1-safe design with the primary copies in some central location as these entities may be common for many different Operators.

However, it should also be possible for Transport operators to create the Locations that are special for them as for example the addresses of the transport Customers and their Customers/Suppliers.

In the planning phase of an Order, it is possible for the order entry ERP system to suggest the Collies for the order and create the corresponding Collie records and Container-route reservations. However, these planned Collies and Container-route reservations may be changed in any location in order to make the planned transports correspond to how the transports take place in the real world. It does not matter whether the transports are operated by a foreign Operator as long as the information about changes is exchanged by using the recommended replication methods.

Example:
Suppose the Container-route reservations are replicated with the 0-safe design with primary copy commit in the ERP system of the Operators that are responsible for Container-route reservation transports in current operation. By using the reread countermeasure in the primary copy location it is possible to prevent the lost update anomaly. If a Container-route reservation for a Package/Collie/Container the change must first be committed locally by compensatable subtransaction in the ERP location of the responsible owner of the old scheduled route of the Container-route reservation. By also making a compensatable reservation for the new Container-route reservation the dirty read anomaly is prevented by the pessimistic view countermeasure. A pivot subtransaction can now commit the global change. Finally, retriable subtransactions replicate the changed Container-route reservations to the secondary copy locations. The "non-repeatable read anomaly" and "the phantom anomaly" are acceptable, as all users will understand that if some information is changed, it is because the system has received more up-to-date information, and that is only good [11].

5   Conclusions and Future Research

In this paper we have described an architecture of an integrated logistics management and control system. We have described how it is possible to use different types of replication designs to integrate the mobile/de-central functionalities with the functionalities of a central ERP system and the logistics systems of cooperating transport sub-suppliers. It has been a design objective that the different local systems may be heterogeneous as different transport operators may not have the same type logistics management system. It has also been a design objective that it should be possible for both mobile and stationary logistics systems to operate in disconnected mode in order to increase the availability of the total distributed system.

We believe that by using the replication designs described in this paper it is in principle possible to change all types of multiuser systems to mobile applications integrated with a central system in such a way that the use of mobile databases will increase the availability of the mobile users and also make it possible to operate in disconnected mode. However, if application code from the from the central system is reused, it is important to be aware that the traditional ACID properties of a central database cannot be reused without change as only relaxed ACID properties are recommended across the boundaries of the mobile/de-central units. On the other hand, the architecture's use of SOA services makes it also possible to integrate heterogeneous ERP modules developed by different software suppliers.

In coming papers, we hope to present many more ERP functionalities that can either benefit from being executed in mobile/de-central ERP modules or be integrated in other ways in order to increase the flexibility of the existing ERP systems. We believe that using different types of replication designs is the most economical way to implement such new mobile/de-central functionalities. In the future, we will therefore also analyze how it is possible to make standardized replication implementations or middleware that can implement the use of the different replication designs by using Meta data descriptions of the common data to make automatic asynchronous updates.

Acknowledgments. This research is partly funded by the 'Danish Foundation of Advanced Technology Research' as part of the 'Third Generation Enterprise Resource Planning Systems' (3gERP) project, a collaborative project between Department of Informatics at Copenhagen Business School, Department of Computer Science at Copenhagen University, and Microsoft Business Systems.

References

1. Berenson, H., Bernstein, P., Gray, J., Melton, J., O'Neil, E., O'Neil, P.: A Critique of ANSI SQL Isolation Levels. In: Proc. ACM SIGMOD Conf., pp. 1–10 (1995)
2. Breibart, Y., Garcia-Molina, H., Silberschatz, A.: Overview of Multidatabase Transaction Management. VLDB Journal 2, 181–239 (1992)
3. Coratella, A., Hirsch, R., Rodriguez, E.: A Framework for Analyzing Mobile Transaction Models. In: Siau, K. (ed.) Advanced Topics in Database Research, July 2002, vol. 2, IGI Global (2002),
   http://www.igi-global.com/bookstore/chapter.aspx?titleid=4349
4. Frank, L.: Replication Methods and Their Properties. In: Rivero, L.C., Doorn, J.H., Ferraggine, V.E. (eds.) Encyclopedia of Database Technologies and Applications. Idea Group Inc. (2005)
5. Frank, L.: Architecture for Mobile ERP Systems. In: Proc. of the 7th International Conference on Applications and Principles of Information Science (APIS 2008), pp. 412–415 (2008)
6. Frank, L.: Architecture for Integrated Mobile Logistics Management and Control. In: Proc. of the 2nd International Conference on Information Technology Convergence and Services (ITCS 2010). IEEE Computer Society (2010)
7. Frank, L.: Design of Distributed Integrated Heterogeneous or Mobile Databases, pp. 1–157. LAP LAMBERT Academic Publishing AG & Co., KG, Germany (2010b) ISBN 978-3-8383-4426-3
8. Frank, L.: Architecture for Integrating Heterogeneous Distributed Databases Using Supplier Integrated E-Commerce Systems as an Example. In: Proc. of the International Conference on Computer and Management (CAMAN 2011), May, Wuhan, China (2011)
9. Frank, L.: Countermeasures against Consistency Anomalies in Distributed Integrated Databases with Relaxed ACID Properties. In: Proceeding of Innovations in Information Technology (Innovations 2011), Abu Dhabi, UAE (2011b)

10. Frank, L.: Architecture for Integrated Mobile Calendar Systems. In: Senthil Kumar, A.V., Rahman, H. (eds.) Mobile Computing Techniques in Emerging Market: Systems, Applications and Services. IGI Global (2011c)
11. Frank, L., Zahle, T.: Semantic ACID Properties in Multidatabases Using Remote Procedure Calls and Update Propagations. Software - Practice & Experience 28, 77–98 (1998)
12. Garcia-Molina, H., Salem, K.: Sagas. In: ACM SIGMOD Conf., pp. 249–259 (1987)
13. Giaglis, G.M., Minis, I., Tatarakis, A., Zeimpekis, V.: Minimizing logistics risk through real-time vehicle routing and mobile technologies: Research to date and future trends. International Journal of Physical Distribution & Logistics Management 34(9), 749–764 (2004), doi:10.1108/09600030410567504
14. Gray, J., Reuter, A.: Transaction Processing. Morgan Kaufman (1993)
15. Johansson, B., Sudzina, F.: ERP systems and open source: an initial review and some implications for SMEs. Journal of Enterprise Information Management 21(6), 649–658 (2008), doi:10.1108/17410390810911230
16. Huebsch, R., Hellerstein, J.M., Boon, N.L., Loo, T., Shenker, S., Stoica, I.: Querying the internet with PIER. In: Proc. of VLDB (2003)
17. Härder, T.: DBMS Architecture–New Challenges Ahead. Datenbank-Spektrum 14, 38–48 (2005)
18. Kurbel, K., Jankowska, A.M., Dabkowski, A.: Architecture for Multi-Channel Enterprise Resource Planning System. In: Krogstie, J., Kautz, K., Allen, D. (eds.) Mobile Information Systems II, vol. 191, pp. 245–259. Springer, New York (2005), http://www.springerlink.com/content/6717916mw6081610/
19. Linthicum, D.: Next generation application integration. Addison-Wesley (2004)
20. Loukopoulos, T., Ahmad, I., Papadias, D.: An overview of data replication on the Internet. In: International Symposium on Parallel Architectures, Algorithms and Networks, I-SPAN 2002. Proceedings, pp. 27–32. IEEE (2002), doi:10.1109/ISPAN.2002.1004257
21. Mehrotra, S., Rastogi, R., Korth, H., Silberschatz, A.: A transaction model for multi-database systems. In: Proc. International Conference on Distributed Computing Systems, pp. 56–63 (1992)
22. Weikum, G., Schek, H.: Concepts and Applications of Multilevel Transactions and Open Nested Transactions. In: Elmagarmid, A. (ed.) Database Transaction Models for Advanced Applications, pp. 515–553. Morgan Kaufmann (1992)
23. Murthy, V.K.: Transactional workflow paradigm. In: Proceedings of the 1998 ACM Symposium on Applied Computing, SAC 1998, Atlanta, Georgia, United States, pp. 424–432 (1998), http://dl.acm.org/citation.cfm?id=330851, doi:10.1145/330560.330851
24. Rao, S., Goldsby, T.J.: Supply chain risks: a review and typology. The International Journal of Logistics Management 20(1), 97–123 (2009), doi:10.1108/09574090910954864
25. Serrano-Alvarado, P., Roncancio, C., Adiba, M.: A Survey of Mobile Transaction. Distributed and Parallel Databases 16(2), 193–230 (2004), doi:10.1023/B:DAPD.0000028552.69032.f9
26. Serrano, N., Sarriegi, J.M.: Open Source Software ERPs: A New Alternative for an Old Need. In: IEEE Software (2006)
27. Sheth, A.P., Larson, J.A.: Federated database systems for managing distributed, heterogeneous, and autonomous databases. ACM Computing Surveys 22(3), 183–236 (1990), doi:10.1145/96602.96604
28. Stockinger, H.: Data Replication in Distributed Database Systems. In: CMS Note (1999), http://cdsweb.cern.ch/record/687136

29. Stonebraker, M., Aoki, P.M., Litwin, W., Pfeffer, A., Sah, A., Sidell, J., Staelin, C., Mariposa, A.Y.: A wide-area distributed database system. VLDB Journal 5(1), 48–63 (1996)
30. Weikum, G., Schek, H.: Concepts and Applications of Multilevel Transactions and Open Nested Transactions. In: Elmagarmid, A. (ed.) Database Transaction Models for Advanced Applications, pp. 515–553. Morgan Kaufmann (1992)
31. Zhang, A., Nodine, M., Bhargava, B., Bukhres, O.: Ensuring Relaxed Atomicity for Flexible Transactions in Multidatabase Systems. In: Proc. ACM SIGMOD Conf., pp. 67–78 (1994)
32. Gligor, V.D., Popescu-Zeletin, R.: Transaction management in distributed heterogeneous database management systems. Inf. Syst. 11(4), 287–297 (1986)

Author Index